Creative Physics Problems

Vol. 2: Waves, Electricity & Magnetism, and Optics

Chris McMullen, Ph.D.

Creative Physics Problems
Volume 2: Waves, Electricity & Magnetism, and Optics

Copyright (c) 2008 Chris McMullen

All rights reserved. This includes the right to reproduce any portion of this book in any form. However, teachers who purchase one copy of this book, or borrow one physical copy from a library, may make and distribute photocopies of selected pages for instructional purposes for their own classes only.

www.faculty.lsmsa.edu/CMcMullen

Custom Books

Textbooks / science / physics

ISBN: 1440458510

EAN-13: 9781440458514

Contents

1 Simple Harmonic Motion 4
2 Waves 9
Review 1: Oscillations and Waves 13
3 Electric Field 17
4 DC Circuits 23
Review 2: Electricity 32
5 Magnetic Field 44
6 Faraday's Law 52
Review 3: Magnetism 57
7 Snell's Law 63
8 Mirrors and Lenses 67
9 Interference and Diffraction 71
Review 4: Optics 73
Final Review: Waves, E&M, and Optics 75
Answers to Selected Problems 82

Volume 2: Waves, Electricity & Magnetism, and Optics

1 Simple Harmonic Motion

Note: An asterisk (*) is used to designate a problem that does not have an answer at the back of the book.

Note: Problems near the surface of the earth involve gravitational acceleration equal to 9.81 m/s^2. However, in some of the answers, this value was rounded to 10 m/s^2 – usually, for problems where the numbers would then work out nicely without the need of a calculator.

1. **Burning Coals**. A monkey is dancing on burning coals. He lifts his feet 30 cm off the ground and his feet touch the burning coals 20 times each minute. What are the (A) amplitude, (B) period, (C) frequency, and (D) angular frequency?

*2. **Lava Lamp**. A bowl of hot lava attached to a vertical spring oscillates up-and-down with a maximum displacement of 9 cm from equilibrium and a period of 2.4 seconds. What are the (A) acceleration and (B) speed of the lava lamp when its displacement from equilibrium is 6 cm? At what position(s) are (A) the acceleration and (B) the speed maximum? zero?

3. **Physics Warms the Heart**. Cupid doesn't use a bow and arrow. Rather, he singes warms your heart with a Bunsen burner. The following graph shows the temperature of your heart as a function of time while you engage in lovely physics thoughts.

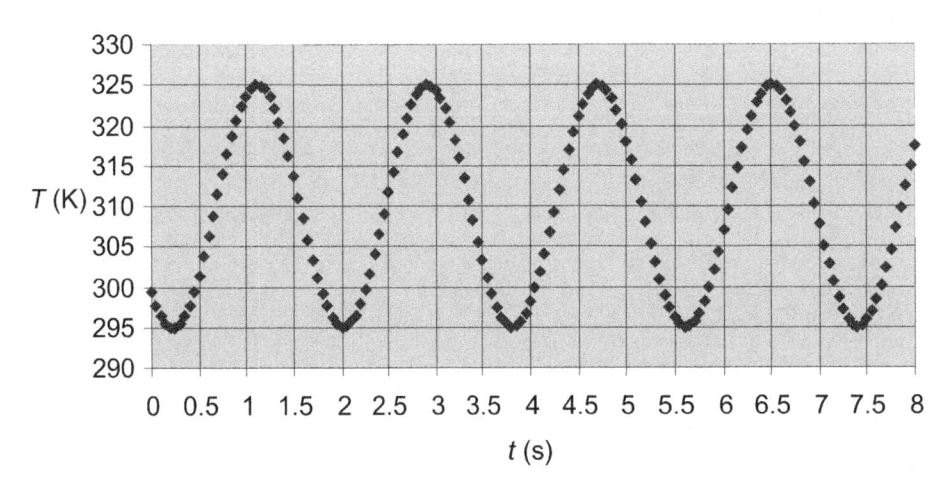

(A) What is the equilibrium temperature? (B) What is the amplitude? (C) What is the period? (D) What is the frequency? (E) What is the angular frequency? (F) What is the phase angle? (G) Write an equation for temperature in terms of time.

*4. **Monkey Bobblehead**. The bobblehead of a monkey oscillates with simple harmonic motion according to the graph below.

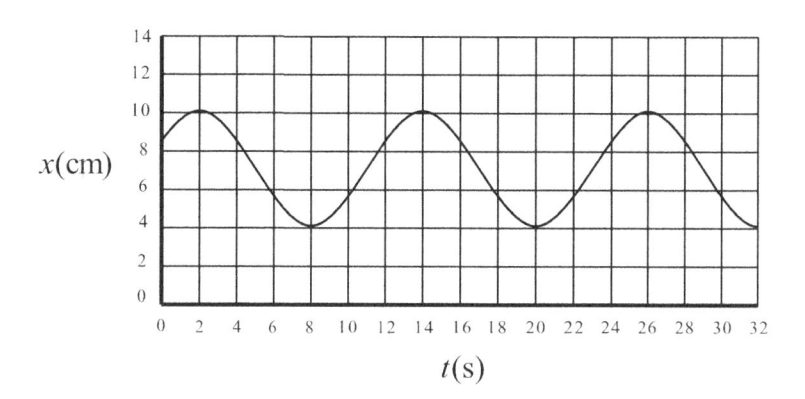

(A) What is the amplitude of oscillation? (B) What is the equilibrium position? (C) How many oscillations will the bobblehead complete in 3 minutes? (D) What is the phase angle? (E) At what time(s) is the bobblehead moving fastest? (E) What is the velocity of the monkey's head at $t = 17$ sec? (F) What is the acceleration of the monkey's head at $t = 26$ sec?

5. **Simple Harmonkey Motion**. A monkey oscillates back-and-forth according to the equation,
$$x = 3\sin(0.4t - 0.6)$$
where SI units have been suppressed. (A) What is the maximum displacement of the monkey from equilibrium? (B) What is the period of oscillation? (C) How far is the monkey *initially* displaced from equilibrium? (D) What is initial speed of the monkey? (E) What is the initial acceleration of the monkey?

6. **Shish Kabob Seconds**. The position of a 50-g shish kabob attached to a spring is
$$x(t) = 0.04\sin(0.3t + 0.7) - 1$$
where SI units have been suppressed. (A) Find the position at three seconds, the equilibrium position, the spring constant, the kinetic energy at three seconds, and the acceleration when the position is -98 cm. (B) Make a graph of *x* as a function of *t*. (C) At what time(s) is the position -102 cm?

7. **Springtime**. You see a physics textbook that is so beautiful that your 30-g eyeballs pop out of their sockets. Fortunately, your eyeballs are attached to 20 N/m springs. Before your eyeballs popped out of their sockets, the springs were compressed 6 cm from equilibrium. Ignore gravity for this problem. (A) With what frequency do your eyeballs oscillate? (B) What is the phase angle? (C) How fast are your eyeballs moving when they are 4 cm from your sockets? (D) What is the acceleration of your eyeballs 4 seconds after popping out of your sockets?

Volume 2: Waves, Electricity & Magnetism, and Optics

*8. **Ice Cube**. A spring with constant 20 N/cm lies horizontally on frictionless ice, with one end attached to an igloo. A 400-g ice cube is pushed against the free end of the spring, compressing the spring 20 cm. The ice cube is then released from rest. (A) Find the acceleration and speed of the ice cube when the displacement of the spring from equilibrium is 12 cm. (B) How much time passes after the ice cube is released before the spring returns to its equilibrium position?

*9. **Penguin Period**. A penguin is suspended from a vertical spring, stretching the spring 12 cm. The penguin is pulled down an additional 6 cm and released from rest. (A) What is the penguin's initial upward acceleration? (B) What is the frequency of the penguin's oscillation? (C) How many oscillations does the penguin complete in one minute?

*10. **Coconut Milkshake**. You mix your 700-g coconut milkshake by attaching it to a 40 N/cm spring and compressing it 8 cm from equilibrium. How fast is the milkshake moving when it is (A) at equilibrium, (B) stretched 3 cm from equilibrium, and (C) stretched 8 cm from equilibrium? (D) Where is the milkshake when it moves with 25% of its maximum speed?

*11. **Lava Rocks**. A clever monkey constructs a lava rock launcher consisting of a cylinder with one end open and a spring inside. When the cylinder is held horizontally at a height of one meter above the ground, a 30-g lava rock is pushed against the spring, compressing it 18 cm. The spring constant is 2 N/m. Once released, what horizontal distance will the lava rock travel before striking the ground?

*12. **Banana SHM**. A bronze banana attached to the free end of a horizontal spring oscillates with simple harmonic motion. The spring constant is 14 N/m, the mass of the banana is 8 kg, and the amplitude of oscillation is 3 m. The displacement of the bronze banana from equilibrium is 2 m, heading in the positive direction, when $t = 0$. (A) What is the displacement from equilibrium when $t = 300$ ms? (B) What is the speed of the banana when $t = 300$ ms? (C) What is the acceleration of the banana when $t = 300$ ms? (D) What is the maximum speed of the banana? (E) Write an equation for the velocity as a function of time.

13. **Cuckoo! Cuckoo!** Einstein's 12-kg head is suspended horizontally from a spring attached to a wall. Einstein's head is pulled 3 cm from equilibrium and released from rest. Every time Einstein reaches his rightmost position he says, *Cuckoo!* You count 45 *Cuckoo!*'s every half-minute. (A) Find the spring constant. (B) Find Einstein's speed and acceleration when the spring is displaced 1 cm from equilibrium.

14. **Good Morning, Physics!** As illustrated below, a 10π-kg monkey sleeps on a 30π-kg bed on frictionless ice. The spring constant is 32π N/cm. The spring is presently stretched 20π-cm from its natural length. This elaborate alarm clock cuts the cord at 3:14 a.m., thereby sending the bed and monkey into simple harmonic oscillation. What a way to start the day!

(A) What is the speed of the bed after traveling 30π cm to the right of its initial position? (Yes, this *is* possible.) (B) How much time elapses before the bed is 30π cm to the right of its initial position? (C) Write down an equation for the acceleration of the bed at time t after the cord is cut. [No variables except for time are permitted in your final answer.] (D) What is the speed of the bed at 3:19 a.m.?

*15. **Feastday**. On Feastday, monkeys who are connoisseurs of fine bananas gather in a large dining hall to enjoy a myriad of delicious recipes ranging from traditional dishes such as the banana-cream pie and banana split to more creative dishes like the banana-stuffed bell pepper. The food is so amazingly wonderful that when their stomachs are filled to the brim, they visit the little monkeys' room to empty their stomachs via simple harmonic regurgitation, as described below.

A 20-kg monkey securely ties himself to one end of a 500 N/m horizontal spring on frictionless ice; the other end of the spring is connected to a wall. A monkey attendant compresses the spring 5 m from equilibrium, then lets go (from rest). The 20-kg monkey then oscillates back and forth with simple harmonic motion.

(A) How many oscillations does the monkey complete in π minutes? (B) Write an equation for the monkey's position as a function of time. Plug in numbers for every symbol except for the two variables. (C) What is the maximum speed of the monkey? (D) What is the monkey's speed when the acceleration is half its maximum value?

16. **Penguilum**. A frustrated student weary of solving animalistic physics problems finds a 30-kg penguin, suspends it from a tree with a 7-m slimy vine, and gives him a great big push to help reduce some of that daunting stress. How many oscillations will the penguin complete in a minute?

17. **Planet Ribbit**. A frog on planet Ribbit attached to the free end of a horizontally oscillating spring with constant 20 N/m has a maximum compression of 12 cm and a maximum speed of 3.6 m/s. When the same frog swings from the end of a 1.7-m vine it has a period of 2.8 seconds. How much does the frog weigh on planet Ribbit?

*18. **Lava Rock Clock**. What should be the length of a lava rock pendulum near the surface of the earth if it is to have a period of exactly one second?

*19. **Moonkey Physics**. A monkey transports a pendulum from the earth to the moon. By what percentage does the period increase or decrease on the moon compared to the earth?

20. **Snow SHO**. A simple pendulum consists of a 50-g snowball connected to an 80-cm string. The snow pendulum is displaced 20º from the vertical and released from rest. (A) Where is the pendulum bob three seconds after release? (B) What is the maximum angular speed of the pendulum bob? (C) What is the maximum angular acceleration of the pendulum bob? (D) Find the tangential speed of the pendulum bob three seconds after release. (E) How much time passes before the tangential acceleration is half its maximum value?

*21. **Pandalum**. A cute, fuzzy panda swings from a long vine. The panda makes a maximum angle of 14º with the vertical and completes 12 oscillations in one minute. Define $t=0$ such that at $t=0$ the panda makes a 5º angle with the vertical and is losing speed. (A) What is the length of the vine? (B) What is the phase angle? (C) What angle does the panda make with the vertical when $t=4\,\text{s}$? (D) What is the maximum angular speed of the panda? (E) What is the tension in the vine when $\theta = 0º$?

*22. **Monkeyball**. An 80-g monkeyball attached to a 16-m long monkeytail is at rest until you punch it, upon which it takes off with an initial horizontal speed of 2 m/s. (A) What is the period of oscillation of the monkeyball? (B) Write an equation for the angular speed of the monkeyball as a function of time. (C) What maximum angle does the monkeyball make with the vertical? (D) If your fist is in contact with the monkeyball for 10 ms during the punch, what is the average force your fist exerts on the monkeyball?

23. **Monkeystick**. A monkeystick is like a meterstick, except that it is 40-cm long (the height of a famous physics monkey, of course). One end of a 3-kg uniform monkeystick is attached to the ceiling via a hinge. A monkey tilts the monkeystick 15º, and releases it from rest. The monkeystick oscillates back and forth with approximately simple harmonic motion. Neglect friction in the hinge. What is the approximate period of oscillation?

2 Waves

1. **Surfer Physics**. You paddle out into the ocean, ready to catch a wave back to shore. While floating you count 28 waves passing underneath you in 3 minutes and 30 seconds. You measure the wave crests to be about 50 meters apart, and the crests to be about 24 meters higher than the troughs. Finally, you catch a wave and surf back to shore. What are the (A) amplitude, (B) wavelength, (C) frequency, and (D) period of these ocean waves? (E) How fast will you be surfing as you ride the wave back to shore?

*2. **Pi in the Sky**. A stream of π's zooms by along a sinusoidal path in follow-the-leader fashion. You count fifty π's passing by each minute, measure the distance between two consecutive troughs to be ten meters, and measure the maximum height to be twenty meters greater than the minimum height. What are the (A) amplitude, (B) wavelength, (C) period, (D) frequency, and (E) speed of this 'π-wave'? [Note: This problem was sponsored by the number π.]

*3. **Swan Lake**. Below the waves of some swans swimming in a pond are drawn <u>to scale</u>: One scaled cm equates to 40 cm. Each swan bobs up and down with simple harmonic motion with a frequency of 5 Hz.

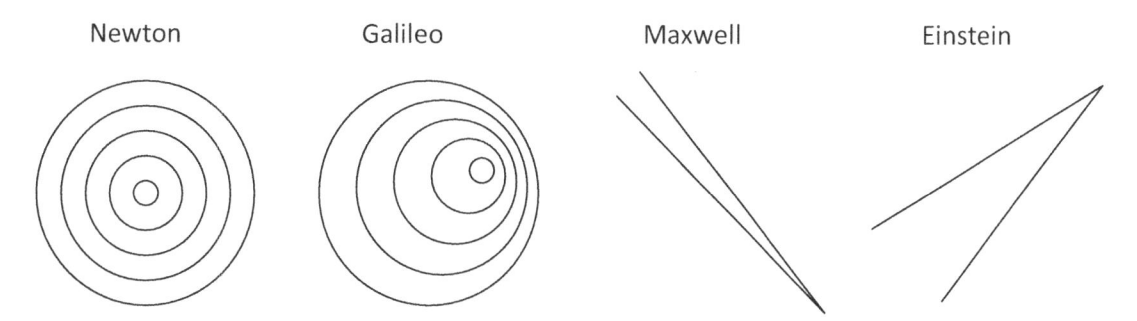

Newton   Galileo   Maxwell   Einstein

(A) List the order of the swans' speeds from greatest to least. Justify your answer. (B) Indicate the direction each swan is swimming. (C) What is the wavelength of Newton's water waves? (D) What is the period of Newton's water waves? (E) What is the speed of Newton's water waves? [**Note**: You may need to use this answer in future questions.] (F) What physics concept best describes Galileo's water waves? (G) If a stationary observer <u>behind</u> Galileo measures the frequency of the water waves behind Galileo, what will be the measured frequency? (H) How fast is Galileo moving? (I) What physics concept best describes Einstein's water waves? (J) How fast is Einstein swimming?

*4. **Rain or Shine**. Rain or shine, the party will continue, which is good because there is a tiny thundercloud off in the distance headed right this way. After seeing a lightning flash you count to seven-one-thousand before hearing thunder. (A) How far away is the storm? (B) Three minutes later you count to two-one-thousand after seeing a lightning flash before hearing thunder. Approximately how quickly is the storm approaching?

Volume 2: Waves, Electricity & Magnetism, and Optics

5. **Dolphin Symphony**. Up in these clouds there is even an ocean where dolphins are heard discussing physics problems underwater. A dolphin produces a sound wave with a frequency of 5 kHz in water. (A) What is the wavelength of this sound wave underwater? (B) The sound wave leaves the water and enters 30º C air. What is the wavelength of this sound wave in the air?

6. **Meet Newton**. Newton wanders aimlessly through physics heaven, followed by an apple tree, as he writes the *Principia*. You can hear him speak the words as he writes them. He wanders through a fog of argon gas. How fast does sound travel in argon gas at room temperature?

7. **Dogarithm**. A dog barks Newton's laws at 78 dB. (A) What is the intensity of this sound wave? (B) What would be the intensity level if the intensity were doubled?

8. **Physics Causes Nightmares!** Exhausted from homework, you take a nap. You dream of Isaac Newton's ghost flying toward you with a stack of physics quizzes that have been graded in blood while you drive away at 40 m/s. You hear Newton shriek. How fast must Newton run for you to hear his shriek with twice the frequency you would hear if you were both at rest?

9. **Doggler Effect**. A dog runs 10 m/s toward an oncoming emergency physics vehicle driving 40 m/s while blaring a 500-Hz siren. By what factor does the frequency heard by the dog change once the vehicle passes the dog (assuming the dog neither changes speed nor direction)?

*10. **Spooky Fog**. Naturally, an eerie fog sets in shortly after the headhunters begin to beat their drums. You get into a canoe and paddle out into the ocean at a rate of 15 m/s while the headhunters pursue you at a rate of 5 m/s. The headhunters begin to screech with an unshifted frequency of 800 Hz, which you perceive to be 775 Hz. What is the speed of sound in this spooky fog?

*11. **SOS**. A supersonic jet is flying horizontally directly overhead when you begin screaming for help. When the sonic boom is heard 10 seconds later you spot the plane at a 60º angle with respect to the horizontal. (A) What is the speed of the plane? (B) What is the altitude of the plane?

12. **Shark Wave**. A supersonic shark travels Mach 2 underwater (this is different from Mach 2 in air), in a hurry to get to physics class on time, of course. The shark is directly underneath an observer in a life raft above, at a depth of 8 km. How long after the shark is directly beneath the life raft does the observer feel the shock wave?

13. **Physics Causes Migraines!** You live in your own world, Thoughtland, where brain waves travel through the sky. Whenever an object in Thoughtland travels faster than the speed of brain waves, a shock wave is generated. When the shock wave reaches you, you feel an intense migraine. While standing on the ground in Thoughtland, you look straight up and see a high-speed physics thought flying horizontally at an altitude of 4 km. Five seconds after being directly overhead, the physics thought is 25º above the horizontal, at which time you feel an intense migraine. What is the speed of brain waves in Thoughtland?

14. **Sonic Doom**. Have no fear: A frustrated physics student is here to destroy the day! The student is flying a supersonic jet directly overhead when he launches a missile. Twenty seconds later you hear the

sonic boom from the jet, and ten seconds after that you hear the sound of the missile fire. What is the speed of the jet?

15. **Frankenstein's Lab**. You walk into a giant physics lab. One end of a string with linear mass density 30 g/m is connected to an oscillator that vibrates up-and-down. The other end of the string passes over a pulley and is kept taut by suspension of a 20-kg dog that meows. The active length of the string is 1.2 m. Draw the first three resonances and find the first three resonance frequencies.

*16. **Monkey Tales**. As illustrated below, a monkey, standing on a spring, oscillates up and down vertically, with small amplitudes (about a centimeter). His tail is glued to a tree (but don't worry, he has a Ph.D. in applied chemistry). The 20-g tail is 8 m long. The tension in the tail is 200 N. (A) What is the speed of the wave along the tail? (B) What are the boundary conditions? (C) Draw the third overtone. (If you draw other standing waves, circle the one you think is the third overtone.) (D) How many cycles correspond to the third overtone? (Your answer *might* be a fraction.) (E) Solve for the wavelength of the third overtone. (F) What is the resonance frequency for the third overtone? (G) If the monkey's mass is 50 kg, what must be the spring constant if the monkey is to oscillate at the resonance frequency corresponding to the third overtone?

*17. **Helium Chamber**. Inside a room filled with helium are some resonance tubes with one end open and one end closed. The tubes are 132 cm in length and 12 cm in diameter. The temperature of the room is 22º C. Draw the first three resonances and find the first three resonance frequencies.

18. **You've Flipped Your Lid!** You've literally flipped your lid. The top of your head opened up and everything inside flew out. Now you're a 1.700-m long hollow pipe filled with a monatomic gas at room temperature with one end open and one end closed. The resonance frequency of the second overtone is 333.1 Hz. What is the gaseous element inside this pipe?

Volume 2: Waves, Electricity & Magnetism, and Optics

*19. **Physics Loves You!** A canary sings *Physics Loves You!* (arguably the greatest song of all time) inside a 7-m long pipe filled with oxygen (16.0 g/mol) at 427ºC with both ends open. (A) What is the speed of sound inside the pipe? (B) Draw the first three resonances. (C) At what frequencies will the canary's lovely tune sound the richest? That is, find the first three resonance frequencies.

*20. **Monkey Clamp**. (A) A monkey clamps a steel rod one-third along its length and produces compressional waves along its length. Can the monkey produce standing waves in this rod? If so, (i) draw the first two standing waves and (ii) express the wavelengths of the fundamental and first overtone in terms of the length of the rod. If not, prove that standing waves cannot be produced in this rod. (B) The clamp is moved to the midpoint of the rod. Repeat (A).

*21. **Monkey Rod**. A monkey sends compressional waves that travel 1600 m/s through a rod with a single clamp 3/8 along its length from one end. Sketch the first overtone and find the corresponding resonance frequency. If this is not possible, prove it.

Review 1: Oscillations and Waves

Swan Lake

Note: The speed of all water waves at Swan Lake is 24 m/s. This applies to Problems 1-6.

*1. **Water Waves**. Consider the wave patterns illustrated below. These are the water waves that were produced by various swans swimming in Swan Lake while kicking their feet with simple harmonic motion.

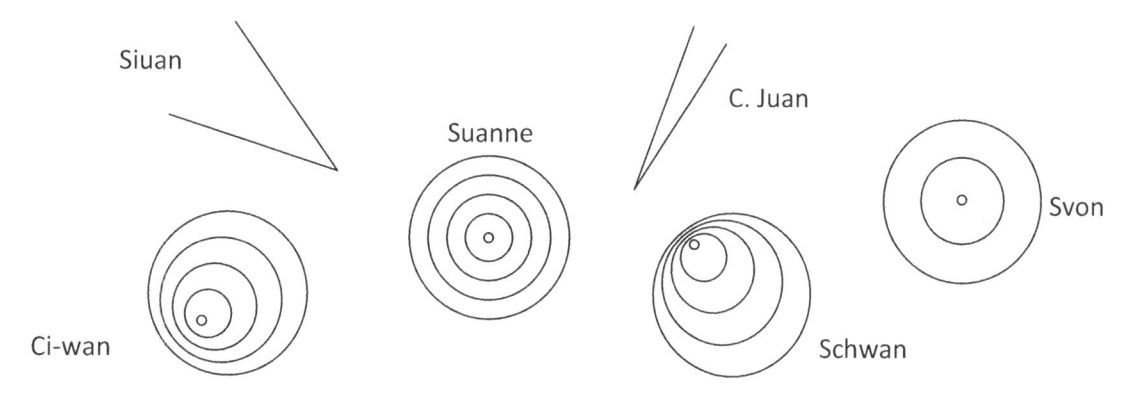

(A) Order the swans' speeds from slowest to fastest. If any have equal speeds, make this clear. (B) Indicate the direction of the velocity of each swan with an arrow. (C) Which physics concept is associated C. Juan? Which physics concept is associated with Ci-wan? (D) If Suanne produces water waves with a frequency of 8 Hz, what is the wavelength of Suanne's water waves? (E) Between Suanne and Svon, who is kicking their feet with the greatest frequency? Explain.

*2. **Hard to Get**. Siouxan is swimming away from Ksihuan with a speed of 18 m/s while producing water waves with a frequency of 12 Hz. Ksihuan pursues Siouxan with a speed of 12 m/s while producing water waves with a frequency of 6 Hz. (A) If Siouxan measures the frequency of the water waves produced by Ksihuan, what will it be? (B) If Ksihuan measures the frequency of the water waves produced by Siouxan, what will it be?

Volume 2: Waves, Electricity & Magnetism, and Optics

*3. **Swim-By**. Tsewan swims by Zwaun with a speed of 48 m/s as illustrated below.

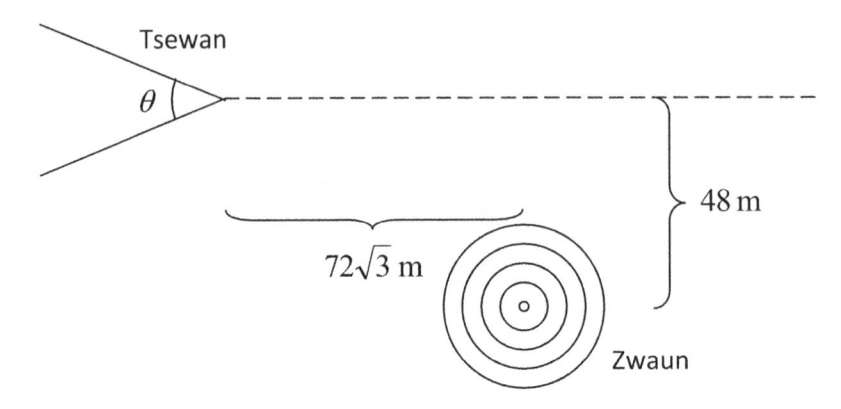

(A) What is the angle θ of the V-shaped wave produced by Tsewan? (B) In the diagram above, draw Tsewan and his V-shaped wave when Zwaun feels the crest of Tsewan's V-shaped wave. (C) From the starting position shown, how far does Tsewan have to swim before Zwaun feels the crest of Tsewan's V-shaped wave?

*4. **Swanusoidal**. Psuan bobs up and down with simple harmonic motion according to the following equation:

$$y = 0.04\sin\left(\frac{3\pi t + \pi}{2}\right) + 0.12$$

where SI units have been suppressed. (However, you must supply correct SI units in your answers.) (A) What is the amplitude of Psuan's oscillation? (B) What is the minimum position of Psuan's oscillation? (C) What is the phase angle? (D) What is the angular frequency of Psuan's oscillation? (E) How many oscillations does Psuan complete in one minute? (F) How fast is Psuan moving at $t = 5\,\text{sec}$? (G) When is Psuan at $y = 10\,\text{cm}$? (H) What is the wavelength of the water waves produced by Psuan?

*5. **Rockabye Baby**. Xiouan connects one end of a 63-N/m spring to a tree on horizontal frictionless ice. Xiouan connects a 7-kg nest of eggs to the free end. Xiouan displaces the nest of eggs 50 cm from equilibrium and releases it from rest. (A) What is the period of oscillation? (B) What is the acceleration of the nest of eggs when it is 25 cm from equilibrium? (C) What is the speed of the nest of eggs when it is 25 cm from equilibrium?

*6. **Mating Calls**. Tse-wan woes Saeioun by making mating calls at one end of a 7-m long pipe filled with oxygen (16.0 g/mol) at 77ºC with both ends open. (A) What is the speed of sound inside the pipe? (B) Draw the second overtone. If you draw additional waves, circle the second overtone. (C) With what frequency should Tse-wan woo Saeioun in order to woo Saeioun with the second overtone?

Dr. McMullen Has a Farm

Dr. McMullen has a farm. $Eie^{i\rho}$! And on his farm he has a cow.
With a μ μ here! And a μ μ there! And a μ μ everywhere!

7. **Cow**. (A) A cow at rest moos with a frequency of 300 Hz. A cow runs 30 m/s in pursuit of a rooster. In this high-speed cow chase, how fast must the rooster run away from the cow in order for the rooster to hear a frequency of 270 Hz? (B) A cow spots a supersonic aircraft directly over its head. The aircraft makes an angle of 40º with respect to the horizontal 8 seconds later when the cow hears the sonic boom. What is the altitude of the aircraft?

Dr. McMullen has a farm. $Eie^{i\rho}$! And on his farm he has a monkey.
With an e^2k e^2k here! And an e^2k e^2k there! And an e^2k e^2k everywhere!

8. **Monkey**. A 70-g monkey is connected to one end of a 40 N/m spring, while the other end of the spring is attached to the bottom interior side of a box. When the lid is attached, the spring is compressed 12 cm from equilibrium. When a child quickly removes the lid (which is on the side, not the top) of this monkey-in-the-box, the monkey pops out, oscillating horizontally with simple harmonic motion. (A) How many oscillations does the monkey complete in one minute? (B) Where is the monkey when his speed is 1.4 m/s? (C) Write an equation for the acceleration of the monkey as a function of time. Replace all other symbols with their numerical values.

Dr. McMullen has a farm. $Eie^{i\rho}$! And on his farm he has an owl.
With a Wh_0 Wh_0 here! And a Wh_0 Wh_0 there! And a Wh_0 Wh_0 everywhere!

9. **Owl**. An owl clamps a 30-cm long, 40-g gummy worm between two posts in order to produce standing waves. The tension in the gummy worm is 50 N. (A) What is the speed of a wave traveling along the gummy worm? (B) Draw the fundamental resonance and the first two overtones. (C) Find the first three resonance frequencies.

Dr. McMullen has a farm. $Eie^{i\rho}$! And on his farm he has a sheep.
With a $\beta\alpha^2$ $\beta\alpha^2$ here! And a $\beta\alpha^2$ $\beta\alpha^2$ there! And a $\beta\alpha^2$ $\beta\alpha^2$ everywhere!

10. **Sheep**. A 40-kg sheep hypnotizes a wolf by swinging from a long vine. The wolf counts 50 oscillations before dozing off 3 minutes later. The sheep makes a maximum angle of 12º with the vertical. Define $t=0$ such that at $t=0$ the sheep makes an angle of –8º and is heading away from equilibrium. (A) What is the length of the vine? (B) What is the phase angle? (C) What is the maximum tangential speed of the sheep?

Dr. McMullen has a farm. Eie^{ip}! And on his farm he has a parrot. With a $\sum \vec{F} = \dfrac{d\vec{p}}{dt}$ here! And a $W = \int \vec{F} \cdot d\vec{s}$ there! And an $E = mc^2$ everywhere!

11. **Parrot.** A parrot carefully studies the oscillatory motion of an enormous 500-g cracker attached to a spring. The position of the cracker is described by
$$x(t) = 4\sin(0.3t - 0.2)$$
where SI units have been suppressed. (A) What is the spring constant? (B) What is the first positive value of t for which the kinetic energy of the cracker equals the potential energy of the spring?

3 Electric Field

*1. **Monkey Equations**. Help a monkey sort through the equations in his physics notes. (A) What formula is useful for computing the force that one charged object exerts on another charged object? (B) What formula is useful for computing the force exerted on a charged particle in the presence of an electric field? (C) If you have already computed the net electric field at point X and then a charge is placed at point X, how can you compute the net force exerted on the charge placed at point X? (D) What formula is useful for computing the potential energy of a charged particle in the presence of an electric field? (E) How is the electric field between the plates of a parallel-plate capacitor related to the potential difference between the plates? (F) What set of equations governs the motion of a charged particle in the presence of an electric field? (G) What equation provides a measure of how well a capacitor stores charge, regardless of its geometry? (H) What equation specifically provides a measure of how well a parallel-plate capacitor stores charge?

*2. **Monkey Analogies**. Help a monkey understand analogies between electricity and gravity. (A) What is the electrical analog of m? (B) What is the electrical analog of g? (C) What is the electrical analog of mg? (D) What is the electrical analog of G? (E) What is the electrical analog of mgh?

*3. **Monkey Strategy**. Help a monkey learn the problem-solving strategies for electricity. (A) Does a single point charge produce a uniform electric field? (B) Does an electric dipole produce a uniform electric field? (C) Is the electric field uniform in the central region between the plates of a parallel-plate capacitor? (D) When there are three charges, outline the general prescription for finding the net force exerted on one of the charges by the other two. (E) When an object is in electrostatic equilibrium, outline the strategy for relating the forces that act on the object. (F) When there are two charges, outline the general prescription for finding the net electric field at any point in space. (G) When a charged particle is moving in the presence of a uniform or non-uniform electric field, outline the general prescription for computing the speed of the charged particle. (H) What strategy would you employ to determine the acceleration of a charged object in the presence of an electric field?

*4. **Monkey Units**. Help a monkey sort out SI units for electricity. (A) What are the SI units of F_E, k, ε_0, q, r, E, ΔU, PE_E, ΔV, d, C, and Q? (B) What do you get when you divide a Joule by a Volt? (C) What do you get when you divide a Coulomb by a Volt? (D) Show that a N/C equals a V/m. (E) Show that a F·J equals a C^2. (F) Show that the units of ε_0 equal a F/m. (G) What do you get when you multiply the units of k with the units of ε_0? (H) Show that a $\sqrt{\dfrac{C \cdot V}{kg}}$ equals a m/s.

5. **Prize-Winning Apples**. Two metal apples have equal mass and positive charge. They attract one another gravitationally, but repel each other electrically. If the net charge on each apple is +2.0 μC, what must be the mass of each apple if the net force exerted on each apple is exactly zero?

6. **Apples and Oranges.** A metal apple and a metal orange are placed 20 cm apart. The apple is charged to 4.0 μC and the orange is charged to −7.0 μC. (A) Find the force between the apple and orange. Is the force attractive or repulsive? (B) The apple is picked up and brought to touch the orange, making electrical contact for a moment. It is then returned to its original position. What is the force between the apple and orange now?

*7. **Magic Bananas.** A poor monkey spends his last nickel on two magic bananas. When he dangles the two 20-g bananas from threads, they "magically" spread apart as shown below. Actually, it's because they have equal electric charge.

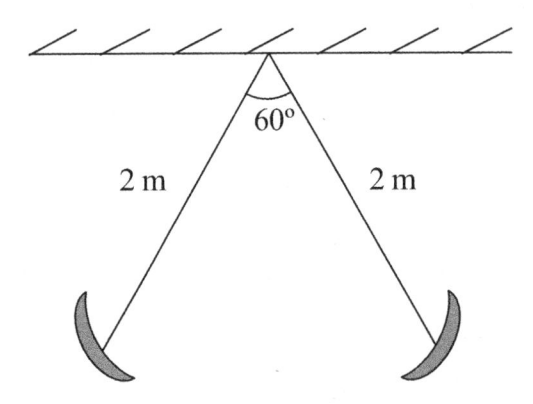

(A) What is the charge of each banana? (B) What is the tension in each string? (C) By what factor would each charge have to increase to make the angle double?

*8. **Just Monkeyin' Around I.** A bored monkeysitter rubs baby monkeys against the carpet in order to charge the baby monkeys. Two charged baby monkeys, which you may assume to be pointlike, are in static equilibrium in the diagram below.

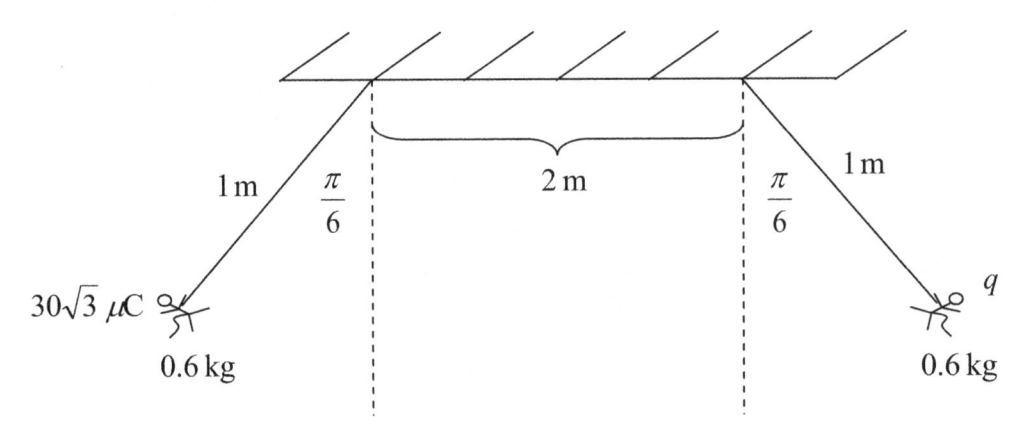

(A) Solve for the tension in each vine. (B) Solve for q.

*9. **Equilateral Triapple**. Three apples form an equilateral triangle. The apples have charges of +2Q, +Q, and −Q. (A) Sketch the lines of force for this charge configuration. (B) Indicate on the sketch regions where the electric field is strongest or weakest. (C) Draw a circle around the +2Q charge. Draw another circle around the +Q charge. Compare the net flux through each circle. (D) Draw a third circle that does not enclose a charge. What is the net flux through this circle?

10. **Triapple**. A metal apple charged to +40 µC is placed at $(-20\,\text{cm}, -40\,\text{cm})$, a metal apple charged to −30 µC is placed at $(50\,\text{cm}, -20\,\text{cm})$, and a metal apple charged to −50 µC is placed at $(20\,\text{cm}, 10\,\text{cm})$. What are the magnitude and direction of the net force exerted on the −50 µC apple?

11. **Apples and Bananas**. A metal apple is charged to 4.0 µC and placed at (20 cm, 20 cm). A metal banana is charged to −7.0 µC and placed at (60 cm, 20 cm). (A) What are the magnitude and direction of the net electric field at (30 cm, 20 cm)? (B) What are the magnitude and direction of the net electric field at (5 cm, 70 cm)? (C) What must x be if the net electric field is zero at $(x, 20\,\text{cm})$? (D) Find the electric potential at (60 cm, 20 cm). (E) Find the electric potential at (30 cm, 20 cm). (F) Consider a metal 'test' mosquito with a positive charge of 30 µC. What would be the net electric force exerted on the mosquito if he were at (5 cm, 70 cm)?

*12. **Bermonkey Triangle**. A monkey blows air into two balloons, twists them to shape them like monkeys, rubs them against his fur to give them a net electric charge, and places them on two corners of an equilateral triangle with 2-m long sides, as illustrated below.

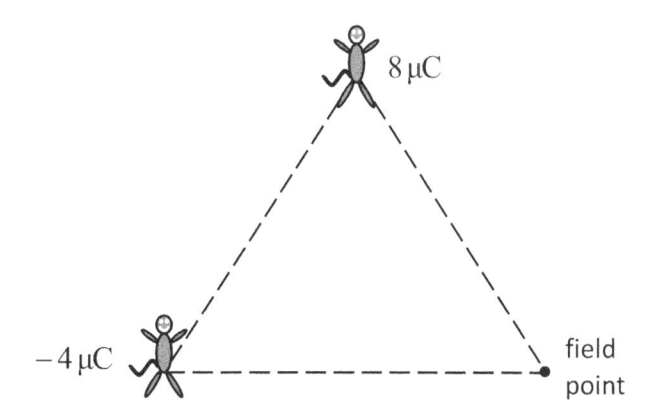

(A) Find the magnitude and direction of the electric field at the bottom right corner of the triangle. (B) What is the electric potential at the field point?

*13. **Just Monkeyin' Around II**. A bored monkeysitter grabs a baby monkey and rubs the baby monkey against the carpet, thereby creating a net charge of $+8$ µC on the monkey. The monkeysitter places the baby monkey at $(1\,\text{m}, \sqrt{3}\,\text{m})$. The monkeysitter similarly charges another baby monkey to -32 µC and places it at $(-2\,\text{m}, 0)$. (A) Find the magnitude and direction of the net electric field at $(2\,\text{m}, 0)$. (B) Find the electric potential at $(2\,\text{m}, 0)$.

14. **A Rose that Froze.** A 40-g metal rose with a net charge of +300 µC is suspended from a tree branch via a thread. The thread makes an angle θ with the vertical, yet the rose is in static equilibrium due to a uniform electric field of 500 N/C at an angle of 30º above the horizontal. (A) What angle does the thread make with the vertical? (B) What is the tension in the thread?

15. **Mosquito Physics.** A metal carrot with a charge of +40 µC is located at (0, 0). A metal head of cabbage with a charge of +80 µC is located at (30 cm, 10 cm). A metal stalk of celery with a charge of −40 µC is located at (20 cm, 50 cm). (A) What is the electric field at (20 cm, 10 cm)? (B) Consider a metal 'test' mosquito with a positive charge of 30 µC. What would be the net electric force exerted on the mosquito if he were at (20 cm, 10 cm)?

*16. **Faraday's Ice Pail.** Faraday's ice pail consists of an inner conducting pail and an outer conducting pail. The outer pail is connected to a ground wire. The pails are initially neutral. Draw two concentric circles to represent the pails. Draw the ground symbol connected to the outer pail. Draw a positively charged sunflower placed inside the inner pail without touching. Draw how the charges on the inner pail redistribute themselves on the inner pail due to the presence of the sunflower. Draw an arrow to indicate the flow of electrons along the ground wire. Draw charges on the outer pail.

17. **Gauss's Claw I.** An infinite, straight monkey tail with uniform charge density lies on the z-axis. Derive an equation for the electric field a distance r_c from the z-axis.

18. **Gauss's Claw II.** An infinite slab – i.e. infinite in two of its dimensions – with thickness d is covered with monkey fur. The monkey fur was rubbed until it had a uniform charge density ρ_0. The coordinates of the slab are:

$$-\frac{d}{2} \leq x \leq \frac{d}{2} \quad , \quad -\infty < y < \infty \quad , \quad -\infty < z < \infty$$

Derive an equation for the electric field inside and outside the slab.

19. **Gauss's Claw III.** An insulating sphere (made from bananas) with a radius R_1 and a total charge $2Q$ has a uniform charge density ρ_0. The insulating sphere is concentric with a spherical conducting shell with an inner radius R_2 and outer radius R_3, where $R_1 < R_2 < R_3$. The conducting shell carries a net charge $3Q$. The system is in electrostatic equilibrium. (A) What is the charge of the inner surface of the conducting shell? (B) What is the charge of the outer surface of the conducting shell? (C) How much charge lies in the range $0 \leq r \leq \frac{R_1}{2}$? (D) Derive an equation for the electric field in each region.

20. **Boring Square.** Three 200-µC charges lie on the corners of a square with 8-cm edges. One corner is therefore empty. Two of the charges are positive, the other is negative. The corner furthest from the empty corner contains a positive charge. Draw this. Find the electric potential at the empty corner.

21. **Ca-pie-citor I.** Two parallel pie pans are connected to a battery, applying a potential difference of 120 V between the pans. A μ^- meson (which has a mass of 139.57018 MeV/c^2 and a charge of −e) leaves

the negative pan heading toward the positive pan with an initial speed of 14 km/s. What is the speed of the μ^- when it strikes the positive pan? [**Notes**: *c* is the speed of light in vacuum, and *e* is the charge of a proton. As always, you can find fundamental constants in your textbook. This is where you can also find the conversion between eV and Joules.]

22. **Ca-pie-citor II**. Two parallel pie pans are connected to a battery, applying a potential difference of 120 V between the pans. The pans are separated by 10 cm. As illustrated below, a pie monkey (which has a mass of 50 g and a charge of -300 μC) leaves the positive pan at an angle of 30º from the positive pan with an initial speed of 40 cm/s. Which pan does the pie monkey strike, and how far does it travel before it strikes the pan? (Just find the horizontal component of the net displacement – not the length of the arc. Assume the pie pans to be infinitely long. Include the effects of gravity.)

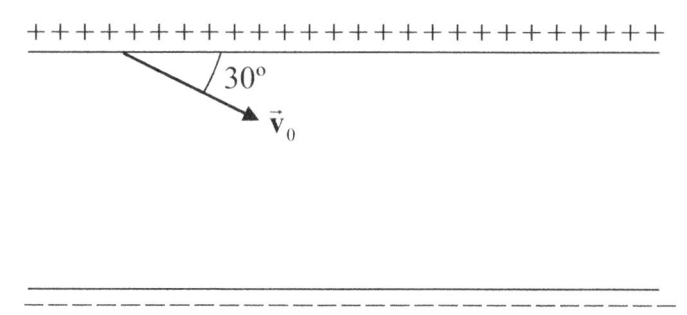

*23. **Phlegmatic Motion**. A sick monkey stands in a uniform electric field, hoping it will make him feel better. A 240-V power supply is connected to two parallel plates separated by 4 m. The monkey's mouth is 1 m above the bottom plate when he coughs. His phlegm, with a mass of 3 g and a charge of -600 μC, has an initial velocity of 4 m/s at an angle of 30º above the horizontal, as illustrated below. Of course, there is gravity, too.

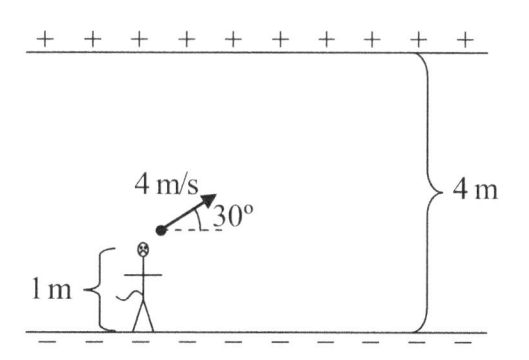

(A) Compute the electric field between the plates. (B) Compute the acceleration of the phlegm. (C) How far does the phlegm travel horizontally before striking a plate, and which plate does it strike? (D) Apply conservation of energy to find the final speed of the phlegm.

*24. **Planet Zharj.**[1] You are kidnapped by aliens (for at least the third time this year, but who's counting?) and transported to Planet Zharj. They dress you in a cute monkey suit, which is specially formulated to hold incredible amounts of charge. They fire you upward at a rate of 30 m/s in a large zero-gravity chamber with an electric field of 160,000 N/C downward. Your mass is 150 kg, and they charged you up to +0.0048 C. (A) What is your acceleration? (B) How high will you rise? (C) How long are you in the air? (D) What is your speed when you are halfway up to your maximum height?

---

[1] Of course, Zharj is pronounced, 'Charge.'

4 DC Circuits

1. **Banana Capacitors**. When a monkey connects two banana capacitors in series, the equivalent capacitance is 12 µF, and when he connects them in parallel, the equivalent capacitance is 50 µF. What are the individual capacitances?

*2. **Fruity Capacitors**. A banana-shaped capacitor has twice the capacitance of an apple-shaped capacitor. When the two capacitors are connected in parallel the equivalent capacitance is 21 µF greater than when the two capacitors are connected in series. What is the capacitance of each capacitor?

3. **Smiley Face**. Of course, the eyes and mouth do not affect anything, as they are not connected to the circuit. Similarly, the hair does not affect anything, as only one end is connected.

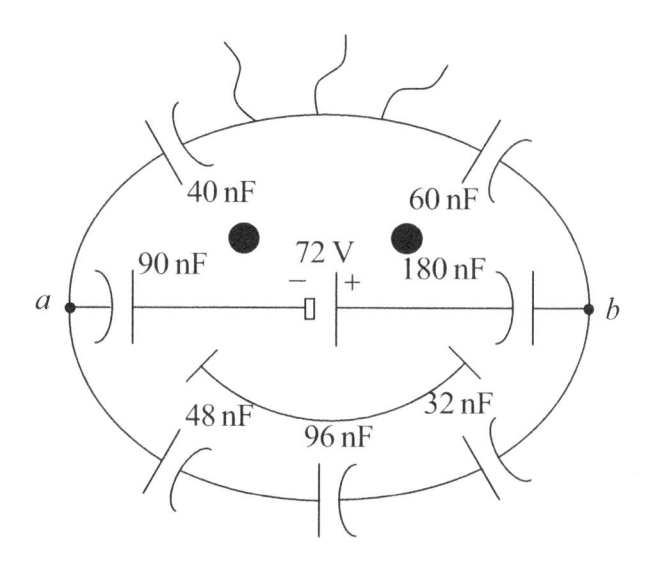

(A) Find the equivalent capacitance of this combination of seven capacitors. (B) Find the potential difference between points a and b. (C) Find the charge stored on the 180 nF capacitor. (D) Find the energy stored by the 60 nF capacitor. (E) Find the charge stored on the 48 nF capacitor.

4. **Home Sweet Home**. Welcome to the house of physics! Consider the house-shaped circuit on the following page. (A) Find the equivalent capacitance of this combination of seven capacitors. (B) Find the potential difference between points a and b. (C) Find the charge stored on the $30\,\mu F$ capacitor. (D) Find the energy stored by the $5\,\mu F$ capacitor. (E) Find the charge stored on the $10\,\mu F$ capacitor.

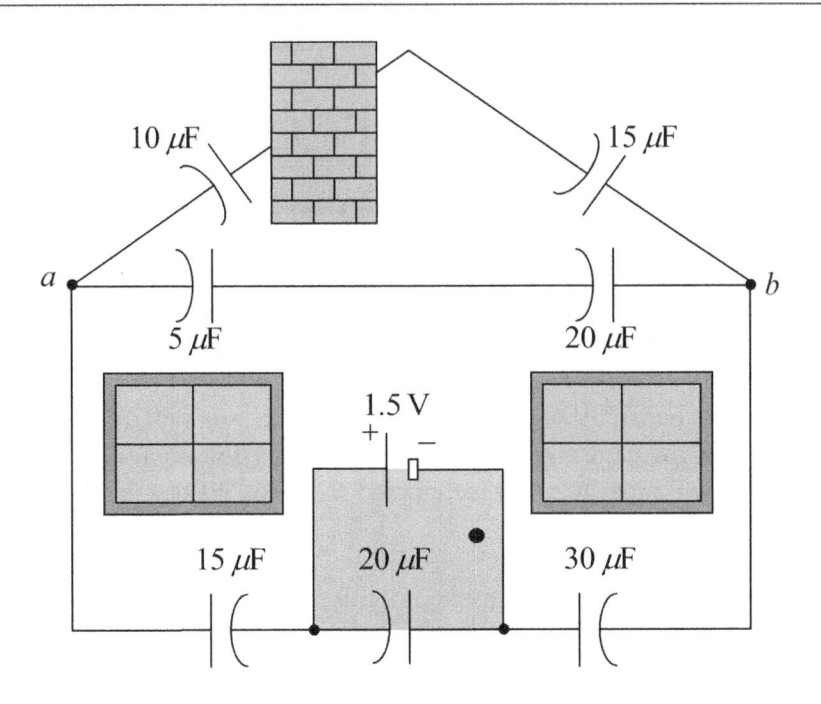

*5. **Bananamobile**. This electric bananamobile is every monkey's dream car.

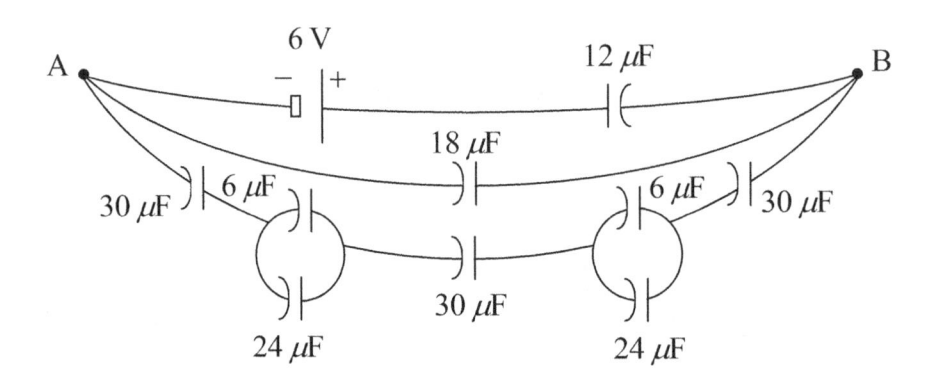

(A) Find the equivalent capacitance of the bananamobile. (B) Find the potential difference between points A and B. (C) Find the energy stored by the 18-µF capacitor.

*6. **Electric Chair**. Sit back and relax in this comfortable shocking chair, illustrated on the following page. This is not three-dimensional; there are no "hidden" wires.

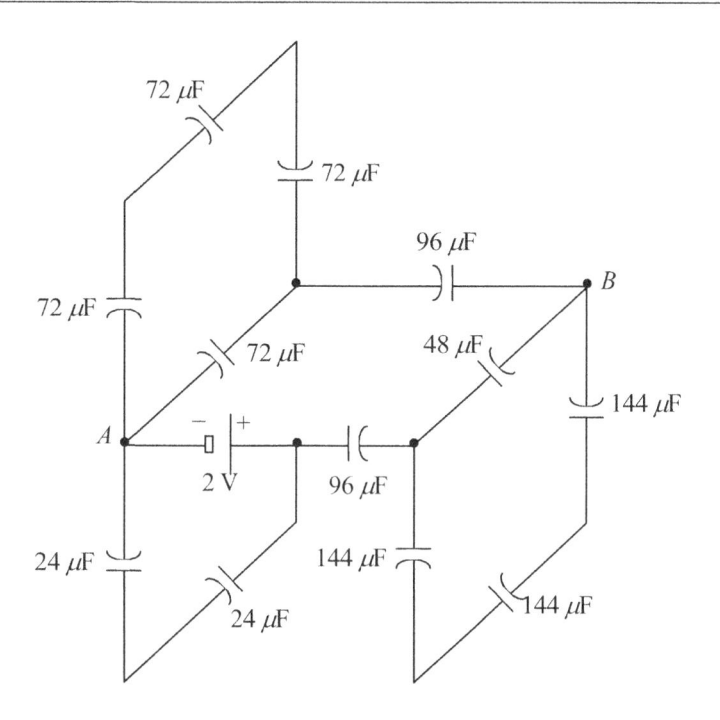

(A) Find the equivalent capacitance of the electric chair. (B) Find the charge stored on each of the 96-µF capacitors. Find the potential difference between points A and B.

*7. **Physics Is a Piece of Cake!** Happy birthday to *you*!

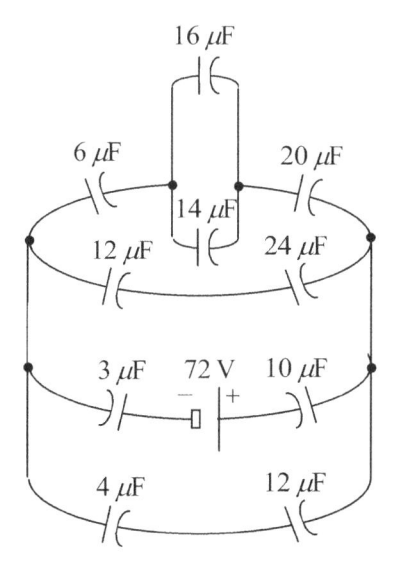

(A) Find the equivalent capacitance of the birthday cake. (B) Find the charge stored on the 20-µF capacitor. (C) Find the energy stored by the 4-µF capacitor.

Volume 2: Waves, Electricity & Magnetism, and Optics

8. **Permit This**. A monkey connects capacitors in a circuit as illustrated below, where $d = 2\,\text{mm}$, $A = 4\,\text{cm}^2$ for the parallel plate capacitors, and the interlocking capacitor has a separation of $1\,\text{mm}$ and effective overlap area of $3\,\text{cm}^2$ per pair of parallel horizontal faces. Dielectric constants: teflon (2.1), nylon (3.4), paper (3.7), quartz (3.8), pyrex (5.6).

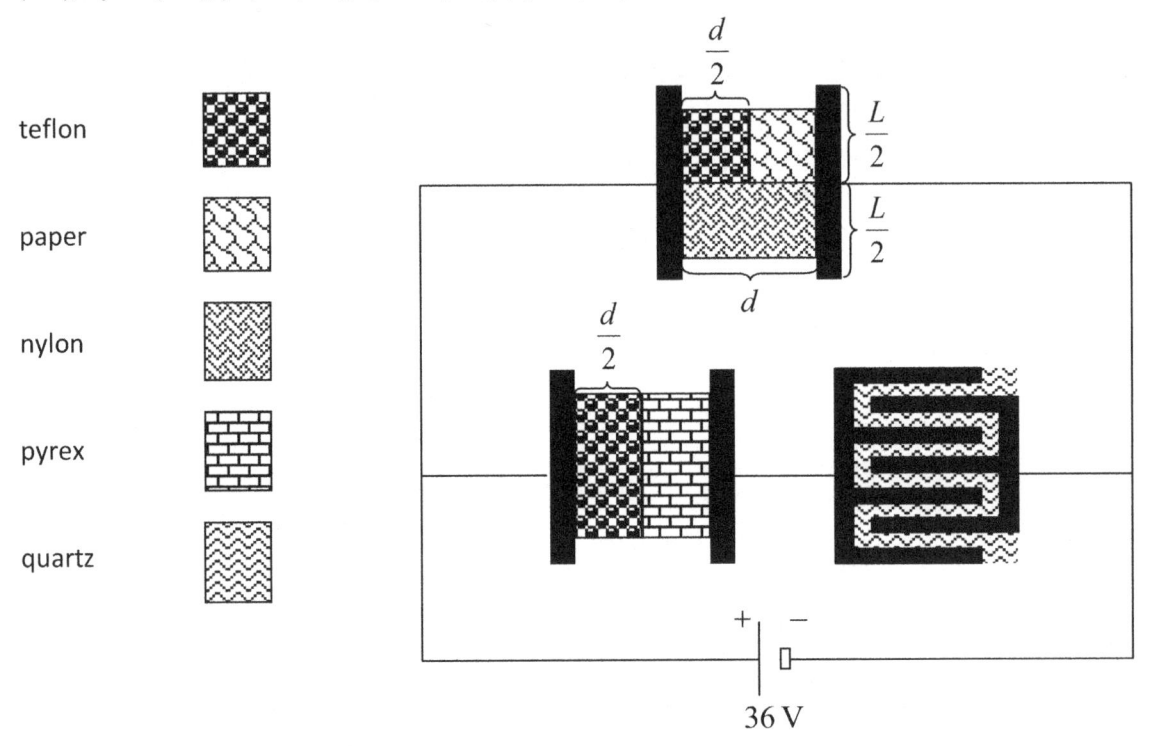

(A) Find the equivalent capacitance. (B) How much energy is stored by the interlocking capacitor? (C) What is the maximum charge that can be stored on the interlocking capacitor? [The rest of the circuit is irrelevant for this part.] Look up any dielectric strengths needed.

9. **Resist This**. A monkey makes a banana-shaped resistor from 100 m of Number 18 aluminum wire. What is its resistance at room temperature? You may need to consult a table of standard wire gauges.

10. **Resistor Journey**. A copper resistor has a resistance of 40 Ω at 20 ºC. You may need to look up properties pertaining to the resistivity of this material. (A) Find its resistance when it makes a journey to a desert where the temperature is 50 ºC. (B) Find its resistance when it makes a journey to a wintery wonderland where the temperature is −10 ºC.

11. **The force that's not a force, of course**. A battery is connected to a variable resistor as shown on the following page. When the variable resistor is set to 100 Ω, the voltmeter reads 12.0 V. The voltmeter reading is 6.0 V when the variable resistor is reduced to 8.33 Ω. Determine (A) the emf and (B) the internal resistance of the battery.

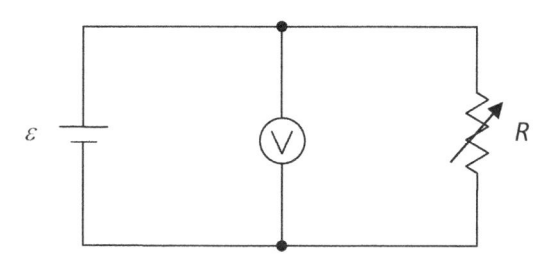

Notation: The arrow through the resistor symbol indicates that it is a variable resistor. A variable resistor is a resistor that can be adjusted to have different values of resistance.

12. **I Scream for Physics!** Three scoops of physics coming right up!

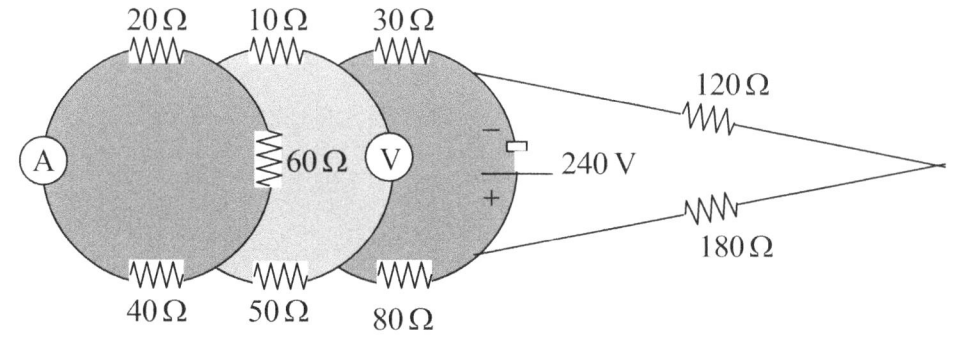

(A) Find the equivalent resistance of this nine-resistor combination. (B) Find the current through the 120-Ω resistor. (C) Find the power dissipated in the 30-Ω resistor. (D) Find the potential difference across the 40-Ω resistor. (E) What does the ammeter read? (F) What does the voltmeter read?

13. **Physics in a Bottle**. A sip of this caffeinated beverage may produce quite a jolt.

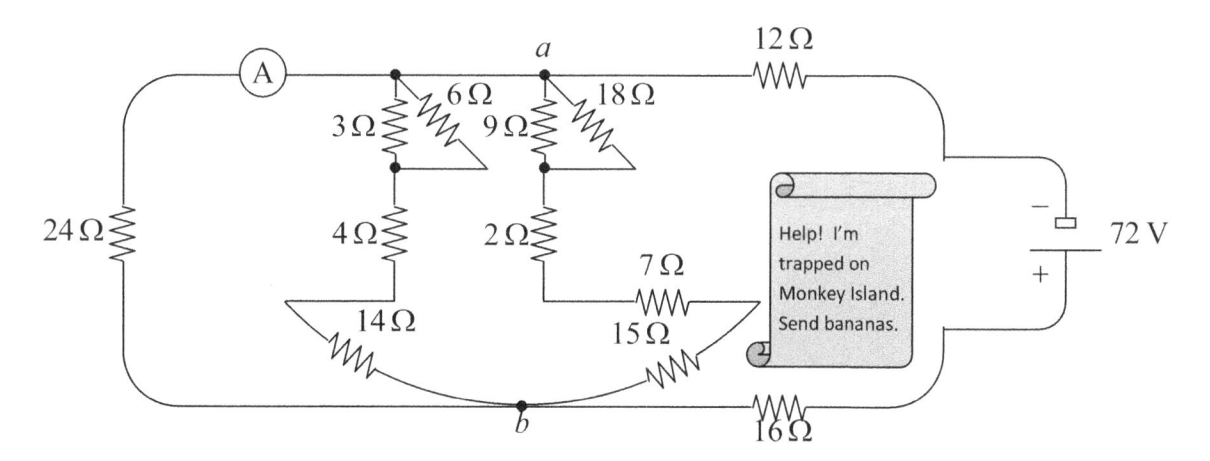

(A) Find the equivalent resistance of this 12-resistor combination. (B) Find the current through the battery. (C) Find the power dissipated in the 3-Ω resistor. (D) What does the ammeter read? (E) What would a voltmeter measure if connected between points a and b?

*14. **Your Brain on Physics.** (A) Find the equivalent resistance for the smiling circuit illustrated below. (B) Find the power dissipated in the 4.0-Ω resistor.

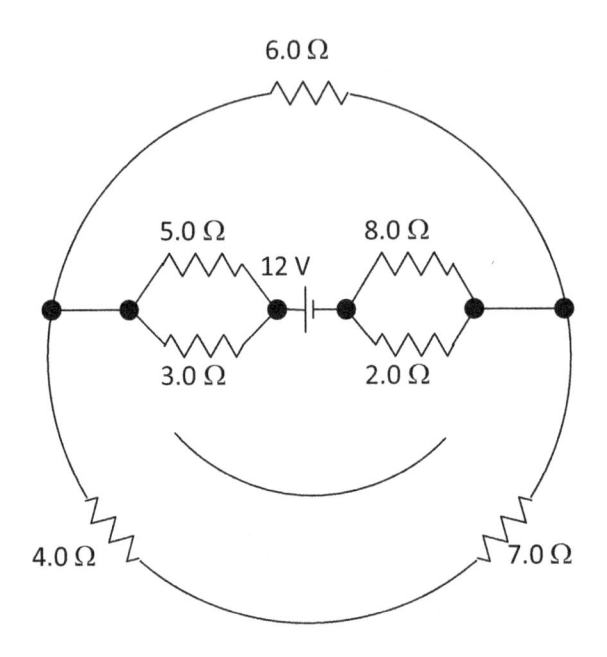

*15. **Bright Idea.** This circuit adds new meaning to *equivalent resistance*.

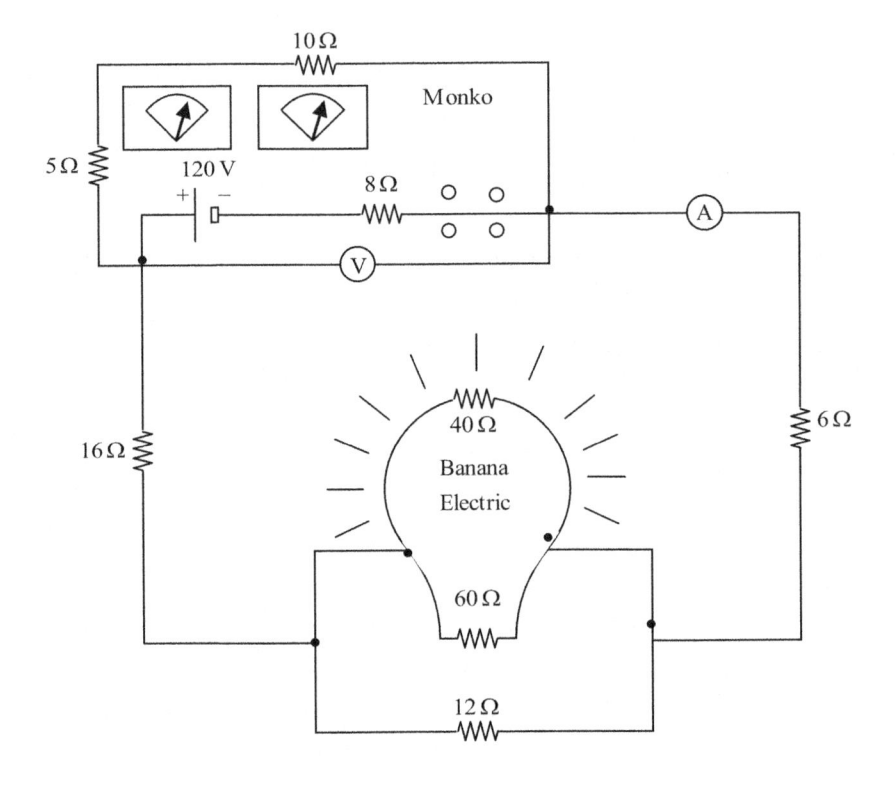

(A) Find the equivalent resistance of the circuit illustrated below. (B) What does the voltmeter measure? (C) What does the ammeter measure?

16. **Sugar Cube**. Twelve identical 1-Ω resistors are connected across the edges of a sugar cube. If a 12-V battery is connected across one of the resistors, what is the equivalent resistance of the twelve-resistor combination?

17. **Resistor Cube**. A battery is connected across the cube of resistors illustrated below. The resistance of each resistor is R. Express the equivalent resistance of the cube in terms of R.

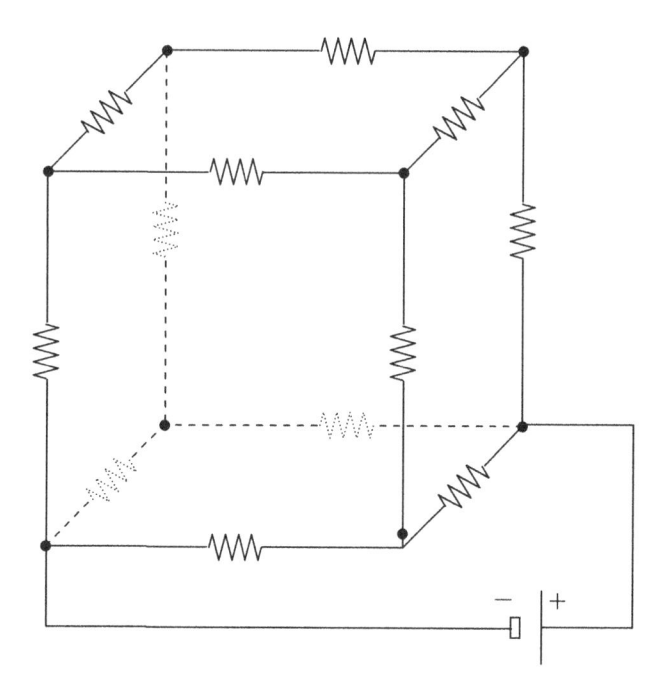

18. **Frosty the Circuit**. Consider the circuit illustrated below.

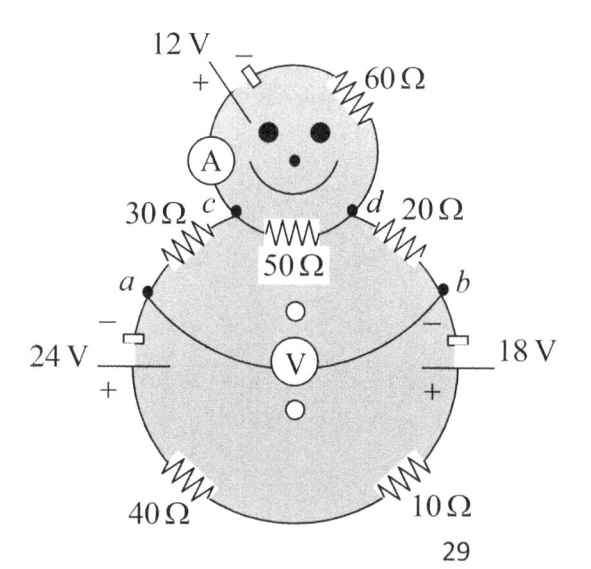

Don't worry about the sign of the ammeter or voltmeter reading. It depends on how you connect the meter. (For example, you aren't told whether the COM of the voltmeter is connected to *a* or *b*.)

Volume 2: Waves, Electricity & Magnetism, and Optics

(A) What does the ammeter measure? (B) What does the voltmeter measure? (C) How much power is dissipated in the 50-Ω resistor? (D) Which of the three currents (top, middle, and bottom) head from d to c? (E) Order points a, b, c, and d, from highest to lowest electric potential.

19. **Monkey Physics**. Consider the circuit illustrated below.

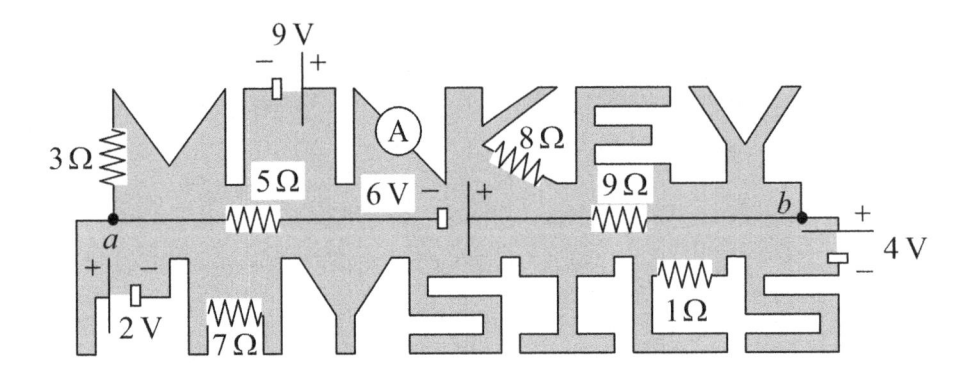

(A) What does the ammeter measure? (B) What is ΔV_{ab}? Sign is important here. $\Delta V_{12} \equiv V_1 - V_2$. (C) How much power is dissipated in the 7-Ω resistor? (D) Which of the three currents (top, middle, and bottom) head from b to a? (E) Of points a, and b, which is at higher electric potential?

*20. **Going Bananas!** Consider the circuit illustrated below.

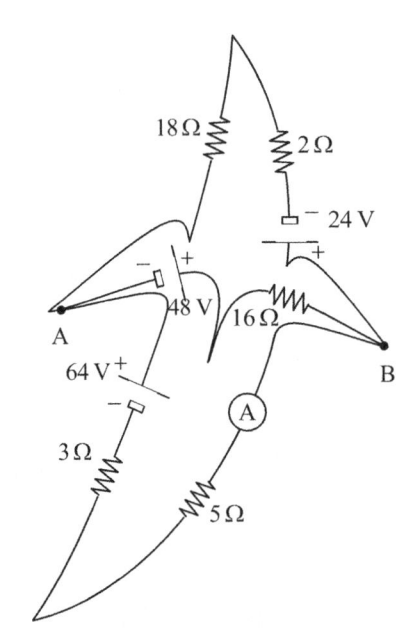

(A) What numerical value, with units, does the ammeter read? (B) Find the potential difference from A to B. (C) If these resistors were lightbulbs, which would glow the brightest? Support your analysis.

*21. **Watterfly**. What do the meters read?

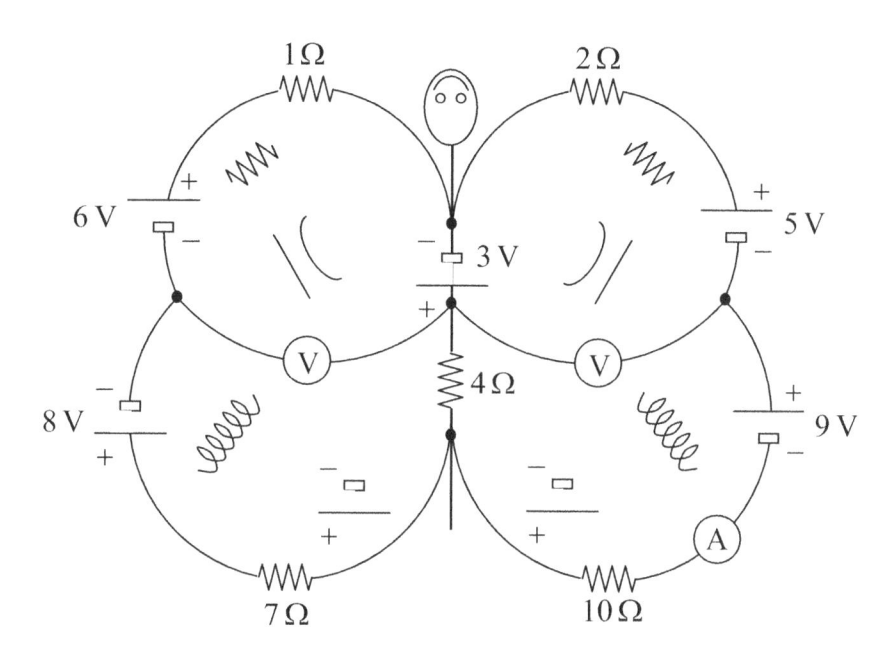

*22. **Coldilocks and the Three Burrs**. Coldilocks is wandering through the snow when she comes across three unoccupied igloos. The first igloo has physics equipment that is too old. The second igloo's physics equipment is too fancy. The third igloo's equipment is just right. Coldilocks decides to use Number 22 nichrome wire to construct a heating coil that delivers 600 W of power when operating on 110 V. (A) What length of wire should be used to deliver the required power at 1000 ºC? (B) Suppose the heating coil is instead heated to 1400 ºC. What will be the actual power delivered by the coil? Use the same length you computed in part (A).

Volume 2: Waves, Electricity & Magnetism, and Optics

Review 2: Electricity

Mr. Principal, I Shrunk my Students

The instructor wheels a large ray gun into class.
Galileo: Hey, everybody! Look at the cool physics equipment!
Students put down their pencils, finding themselves attracted to the front of the room.
Archimedes: I wonder what this button does.
Planck: There's only one way to find out.
Instructor: Be careful not to touch anything.
Einstein presses the shiny red button. The contraption makes a loud hum, followed by a high-pitched whistle. The ray gun begins to rotate as it blasts students with laser beams. After unplugging the device, the instructor looks around the room, seeing nobody.

Maxwell: Now why did you have to go and do that?
Coulomb: If we run around in our socks, we can get charged up and have fun with the equipment.
Kepler: And since we're so tiny, nobody will ever find out!

*1. As illustrated below, Gauss throws Faraday horizontally with an initial speed of 20 cm/s. Faraday has a net charge of 800 μC and a mass of 30 g. The potential difference between the plates is 120 V. The distance between the plates is 50 cm.

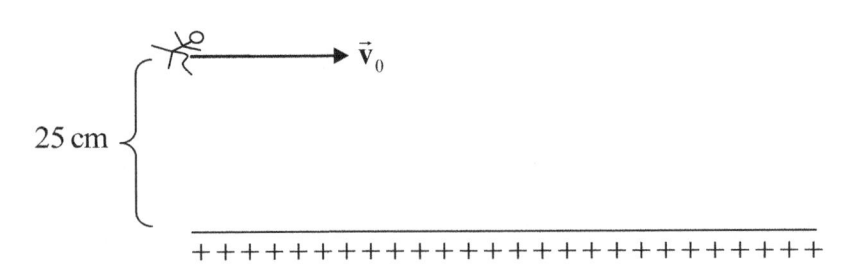

Hertz: Now that's good cooperation between lab partners!
Heisenberg: I wish everyone would stop making useless comments so that I could concentrate…
Michelson: How far does Faraday travel horizontally before striking one of the plates?
Newton: Remember to include the effects of gravity.
Curie: Oh, look! That ray gun gave us all cute little tails!

*2. Bohr has a net charge of -50 µC and stands at (40 cm, 10cm). Rutherford has a net charge of 30 µC and stands at (-30 cm, 20cm).

Schrödinger: What are the magnitude and direction of the net electric field at the origin?

Now Aristotle, who has a net charge of -20 µC, stands at the origin.

Descartes: What are the magnitude and direction of the net force exerted on Aristotle?

*3. Young hangs from a thread suspended from a table above. Young has a mass of 40 g and a charge of 700 µC. A uniform electric field of 600 N/C is directed at an angle of 50º above the horizontal, as illustrated below. The system is in electrostatic equilibrium.

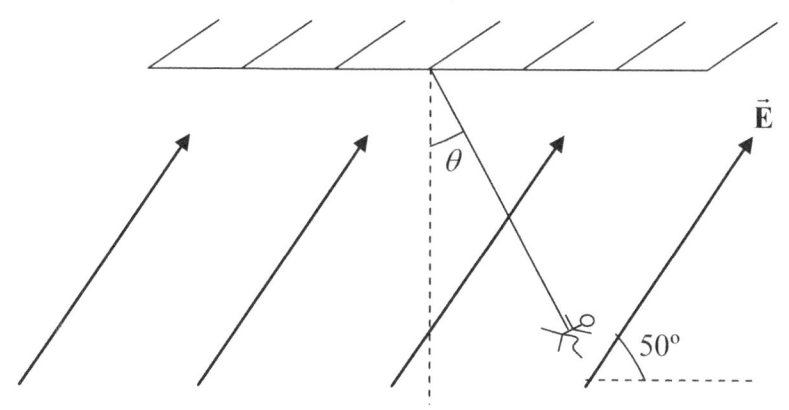

Snell: What is θ?
Cavendish: Obviously, the effects of gravity must be included here.
Young: Oh look! That ray gun gave me one of those cute little tails, too!

Atwood: What is the tension in the thread?

*4. Foucault has a mass of 90 g and a net charge of 700 µC. Foucault jumps straight upward with an initial speed of 20 cm/s in a region where the net electric field is 500 N/C and directed upward.

Lorentz: What is Foucault's acceleration?
Bernoulli: Don't forget to include the effects of gravity.

Feynman: What will be Foucault's maximum height in the air?

Schwinger: How much time elapses before Foucault returns to the ground?

Happy Reseastor!

5. **Reseastor Bunny**. Consider the Easter Bunny circuit below.

(A) Find the equivalent resistance.

(B) What does the ammeter read?

(C) Find the power dissipated in the 14-Ω resistor.

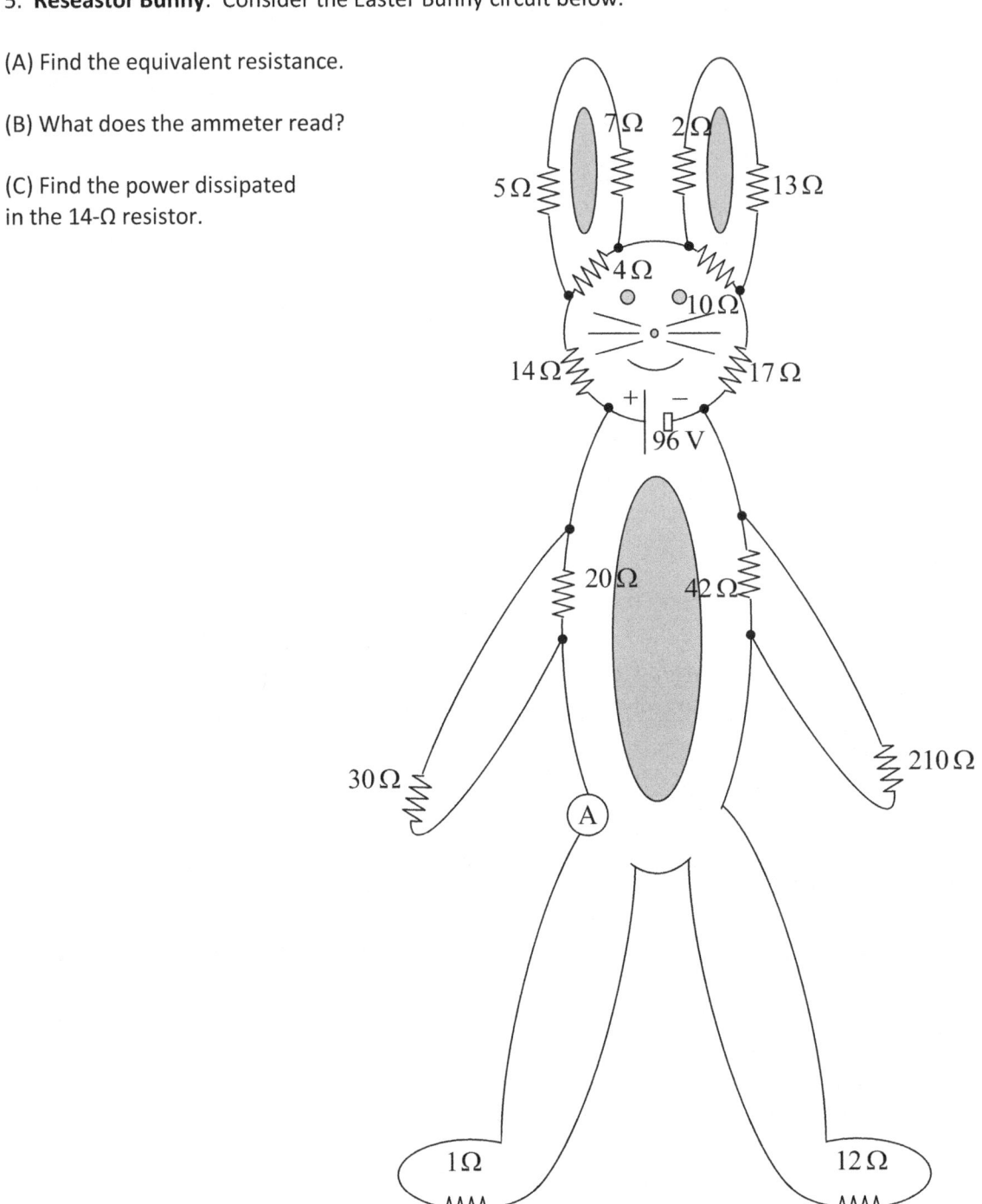

6. **Cabasketor**. Consider the Easter basket circuit below. (A) Find the equivalent capacitance. (B) Find the charge stored on the 6-µF capacitor. (C) Find the energy stored by the 25-µF capacitor.

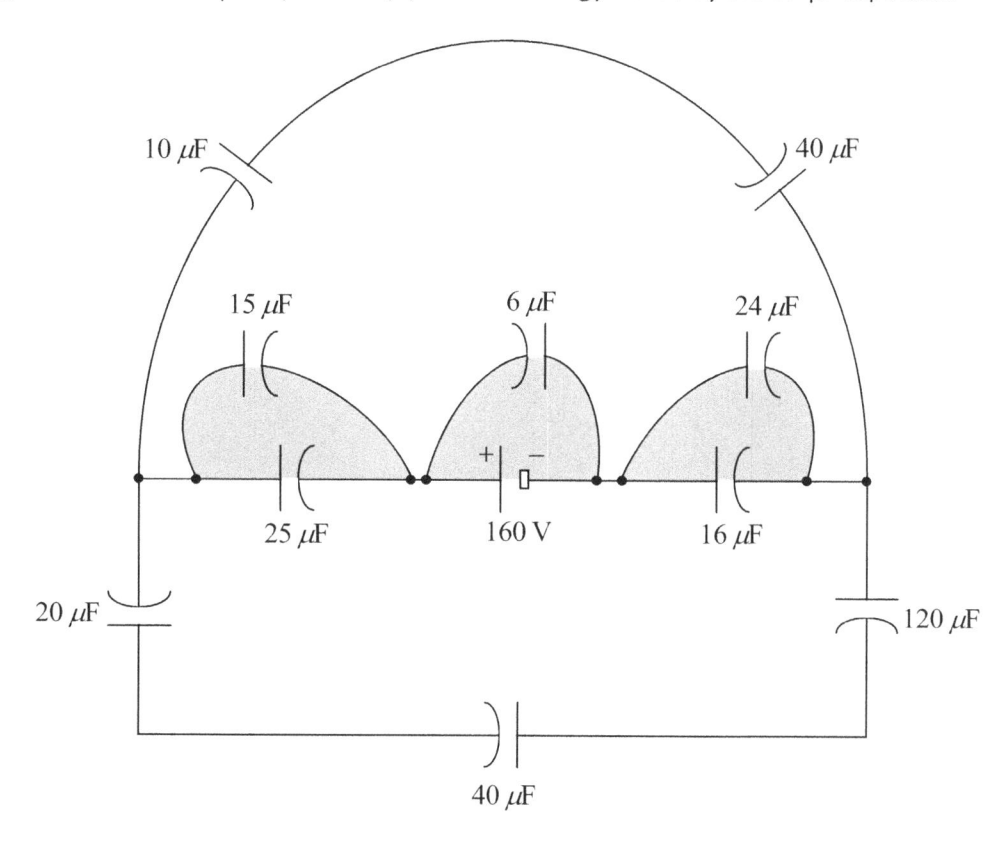

7. **Reseastor Egg**. Consider the Easter egg circuit below. (A) Find the current through the 8-Ω resistor. (B) Find the current through the 12-V battery. (C) What does the voltmeter read?

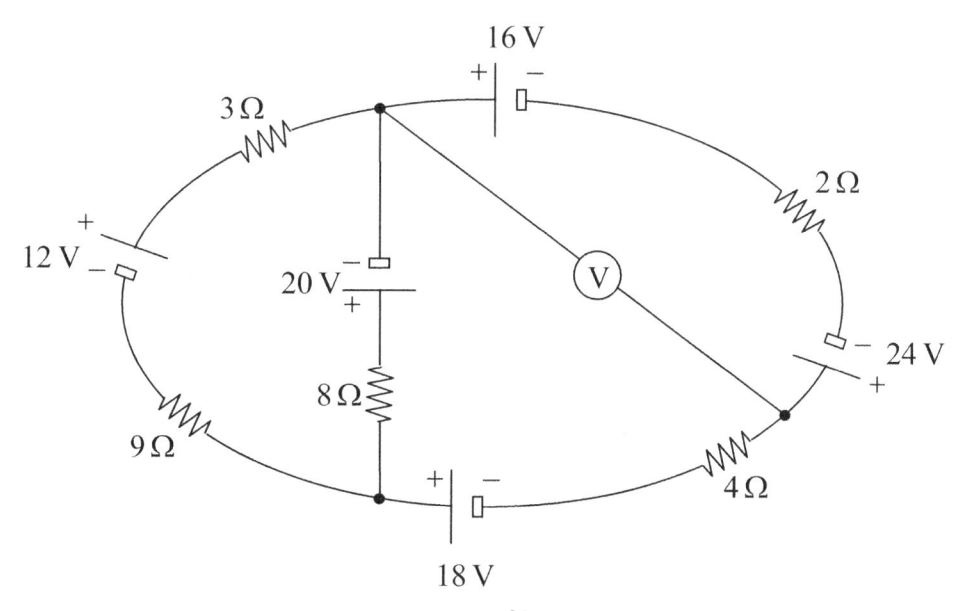

Volume 2: Waves, Electricity & Magnetism, and Optics

Lightning Bros. presents the world-famous

ELECTRIC CIRCUS

The most electrifying show on earth!

*8. **The Big Top**. Find the energy stored by the 4-µF capacitor in the diagram below.

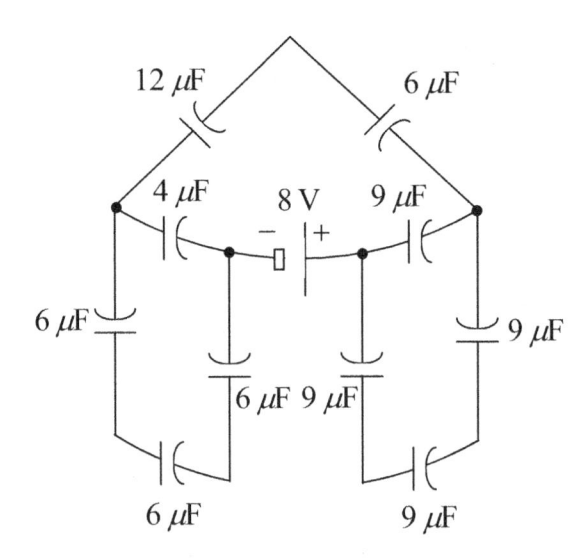

*9. **Cotton Candy**. Find the potential difference between A and B in the diagram below.

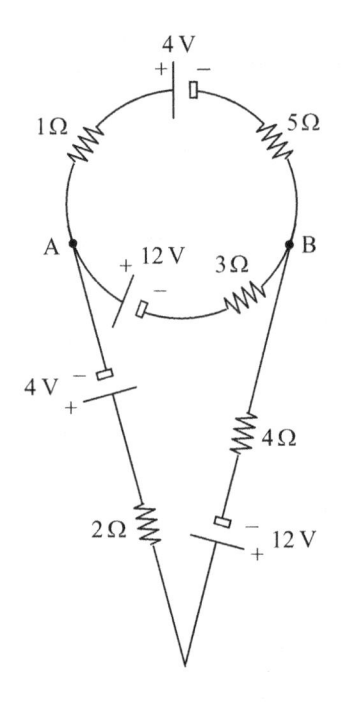

*10. **Juggling**. Find the power dissipated in the 4-Ω resistor in the diagram below.

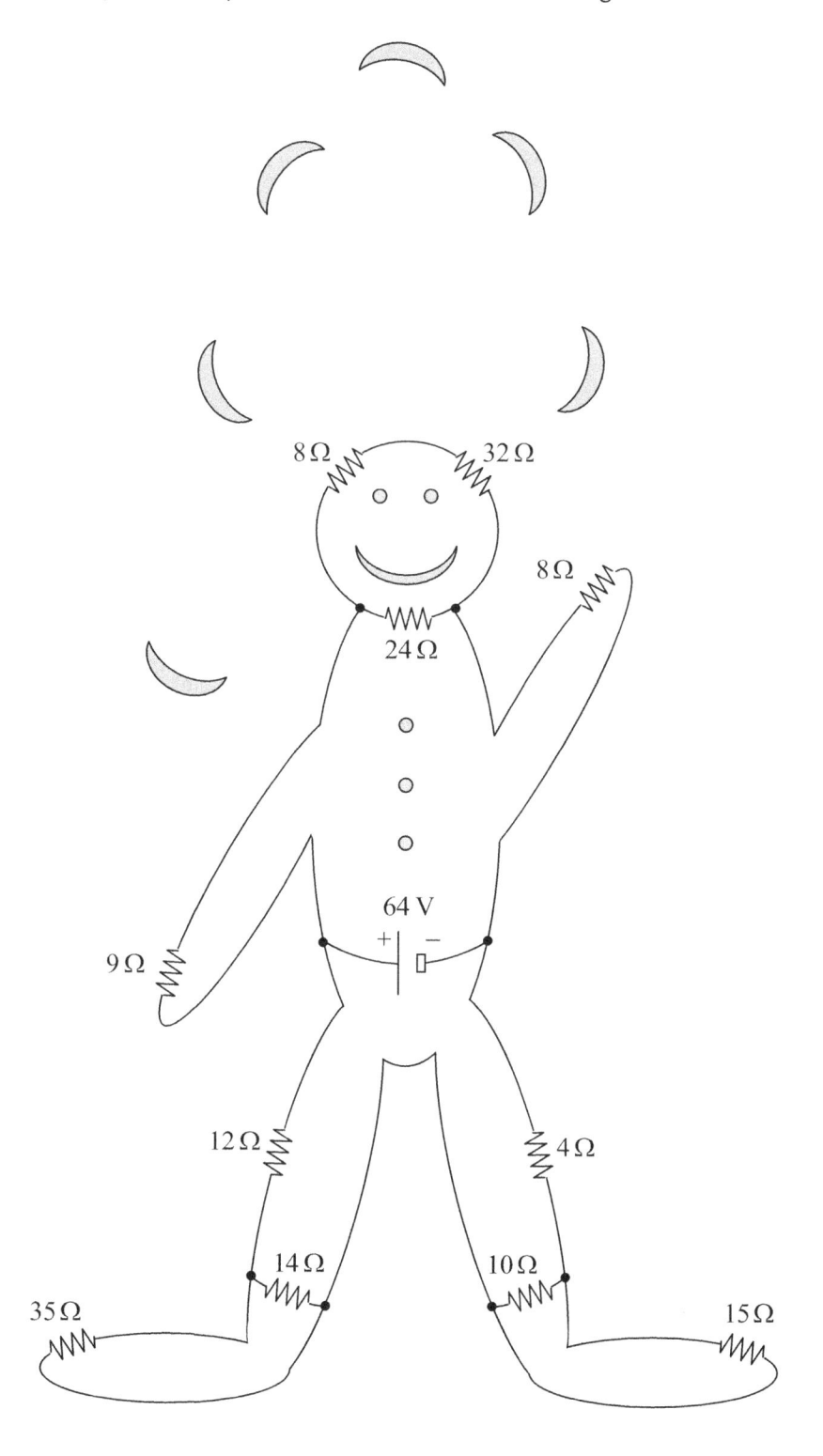

*11. **Lion Cage**. Find the equivalent resistance of the lion cage illustrated below.

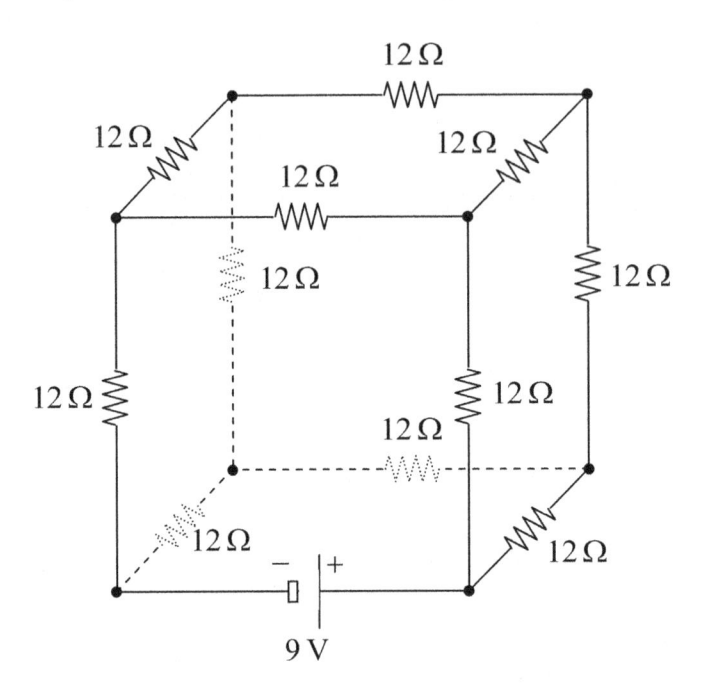

*12. **Flea Circus**. A 200-g flea with a net charge of −200 μC jumps straight upward with an initial speed of $2\sqrt{5}$ m/s in a region where there is a uniform electric field of 3000 N/C directed downward. Remembering to include the effects of gravity, how high does the flea jump?

The Infinitely Long Referential Exam

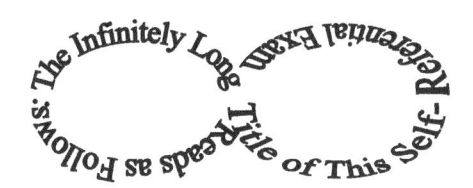

Title of This Self-Referential Exam Reads as Follows:

1. This is the first problem.

This sentence refers to the illustration below, where a banana with a charge of -80 μC lies at $(-3\,\text{m}, 0)$ and a banana with a charge of +40 μC lies at $(3\,\text{m}, 0)$.

Part (A) of this problem asks you to find the magnitude of the electric field at $(0, 3\,\text{m})$.

Part (B) of this problem asks you to find the electric potential at $(0, 3\,\text{m})$.

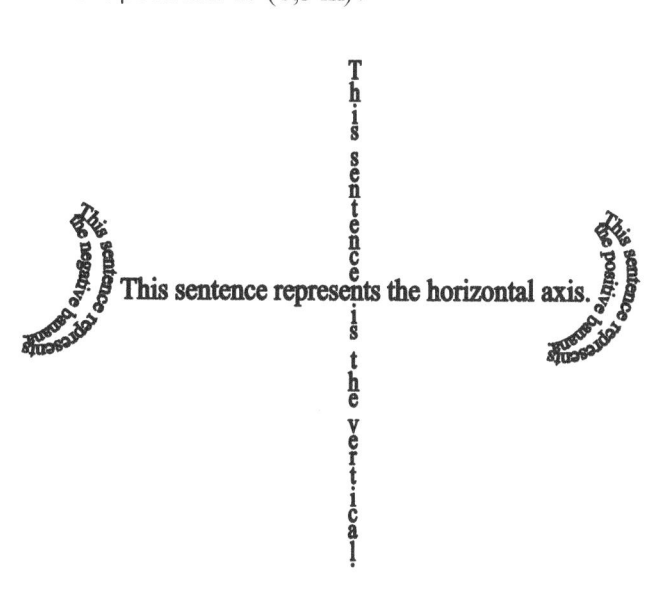

Behold the power of the word!

This sentence commands you to read it.

The negative banana represents This sentence represents the horizontal axis. *This sentence represents the positive banana.*

This sentence is the vertical axis!

2. This is the second problem.

This sentence refers to the diagram on the following page, where a 10-g banana with a charge of -500 μC is initially at rest halfway between two parallel plates separated by 2 m. This sentence states that the potential difference between the plates is 240 V. This statement points out that there is gravity.

Part (A) of this problem asks you to find the electric field between the plates.

Part (B) of this problem asks you to find the acceleration of the banana.

Part (C) of this problem asks you to find the final velocity of the banana.

Volume 2: Waves, Electricity & Magnetism, and Optics

This sentence represents the positively charged plate.

The sentence to the right is incorrect.

The sentence to the left is correct.

This sentence represents the charged banana.

This sentence represents the negatively charged plate.

This is the third problem.

This is a suggestion to sit down and enjoy a shocking banana. (This parenthetical remark points out that the previous sentence is speaking figuratively. Literally!)

Part (A) of this problem asks you to find the equivalent capacitance of the banana.

Part (B) of this problem asks you to find the potential difference between A and B.

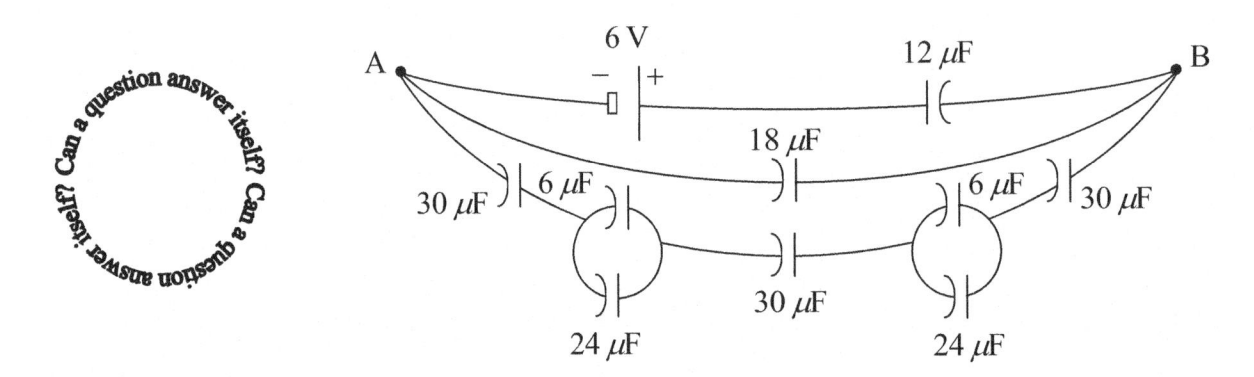

Can a question answer itself? Can a question answer itself?

Circuit: 6 V battery between A and terminal; 12 μF to B; 18 μF; 30 μF; 6 μF; 24 μF; 30 μF; 6 μF; 24 μF; 30 μF.

This is the fourth problem.

This incomplete fragment... this unfinished thought... this redundant tautology...

Part (A) of this problem asks you to draw the electric field lines for a single negative point-charge.

Part (B) of this problem asks to draw the electric field map for the electric dipole (a system of equal and opposite point-charges separated by some distance).

Part (C) of this problem asks you to draw the electric field diagram for a system of two positive point-charges separated by some distance.

40

(This sentence is shy.)
THIS STATEMENT IS BOLD!

Why is this question advising you not to read the first paragraph of this problem? This answer is pointless since you've either already skipped it, in which case you would not be reading this, or you're already reading it, in which case you realized that this answer was a complete waste of time.

Part (A) of this problem asks for the SI units of each of the following:

ε_0 capacitance   electric field

Part (B) of this problem asks which gravitational quantity is analogous to each of the following electrical quantities:

k q E

Part (C) of this problem asks for the conceptual meaning of capacitance.

Part (D) of this problem asks for the three formulas for the energy stored by a capacitor.

Part (E) of this problem asks what happens to the capacitance of a parallel-plate capacitor when you insert a dielectric between its plates.

This is the last sentence.
Correction: That was the last sentence.

Is this question rhetorical? Yes, indeed!

Volume 2: Waves, Electricity & Magnetism, and Optics

The Physics of Love

According to Cool Hombrè's law, there are three kinds of emotional particles in the soul:
- Amourons ♀, which have positive passion p;
- Abhorons ♂, which have negative passion p;
- and apathons ∅, which are romantically neutral.

Cool Hombrè's law is analogous to the inverse-square laws of physics:
- Passion p, which can be positive, negative, or neutral, is the source of a romance field $\vec{R}$.
- Opposites attract, and likes repel.[2]
- Every emotional particle is the source of a romance field $\vec{R}$, analogous to other fields in physics.
- The proportionality constant in Cool Hombrè's law is the romantic constant ♪, which equals 10^6 in SIA units (Système Internationale de l'Amour).
- The SIA unit of passion is the heart ♥.
- Traditional physical quantities, such as force and displacement, retain their usual SI units.

The masses and passions of the fundamental emotional particles are:
$$m_♀ = 800 \text{ mg}$$
$$m_♂ = 400 \text{ mg}$$
$$m_∅ = 600 \text{ mg}$$
$$p_♀ = 2 \text{ μ♥}$$
$$p_♂ = -2 \text{ μ♥}$$

It would be wise to first show that you know the relevant electrical formula before trying to work with the romantic quantities...

$$\langle 1 + 1 = 2 \rangle$$

(A) Express Cool Hombrè's law in symbolic form. (B) Indicate the SIA units of each quantity involved in the above equation. (C) What equation is the romantic analog for the formula for weight? (D) What are the SIA units of the romance field? (E) What is the equation for the romance field of a single point-like emotional particle? (F) How does Cool Hombre's law explain the popular notion regarding the success of long-distance relationships? (G) Sketch the romantic field lines (but **_not_** the equipotentials) for the passion configuration illustrated below. Include arrows to indicate direction.

♂   ♀

---

[2] Contrary to popular belief, this does not imply that opposite sexes are necessarily attracted, nor that same sexes necessarily repel – i.e. the sign of the net passion of one's emotions does not necessarily correspond to one's gender.

$$\langle 2u | \Phi_6 = \infty \rangle$$

(A) Gauss's law for romance is to Cool Hombrè's law as Gauss's law for electricity is to Coulomb's law. Express Gauss's law for romance in symbolic form. (B) What are the SIA units of romantic flux (analogous to electric flux)? (C) Derive an equation for the romance field inside and outside of an infinitely long solid cylinder of radius a with uniform passion density ρ. Include labeled diagrams. (D) What are the SIA units of the passion density ρ?

$$\langle 3 \times | \Phi_6 \, \heartsuit \, U \rangle$$

(A) A monkey with a net passion of 800 m♥ is standing at (-1 m, 0). A monkey with a net passion of −200 m♥ is standing at (2 m, 0). What is the net force exerted on each monkey? Are they attracted to each other? (B) Find the coordinates of the field point where the net romance field equals zero.

$$\langle 4u | t_{\Phi_6 \, HW} > t_{life} \rangle$$

Just as charged particles interact via the electromagnetic field, emotional particles interact via the emotioromantic (ER) field. The romance field is plotted below as a function of position and time for a traveling ER wave. SIA units have been suppressed from the labels of the graph.

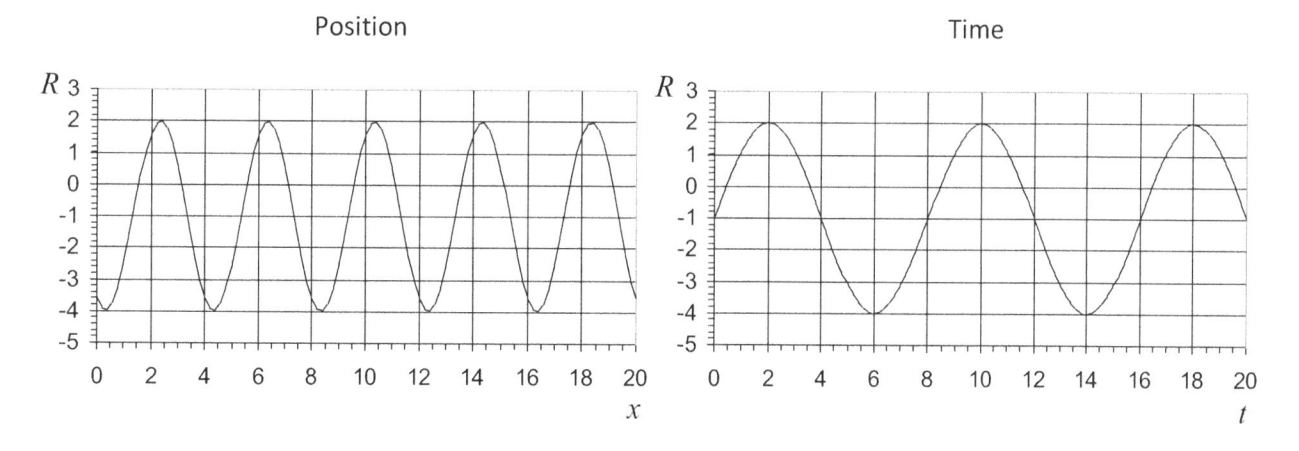

Find each of the following for the ER wave illustrated above (including SIA units, of course): (A) wavelength, (B) frequency, (C) period, (D) angular frequency, (E) k, (F) wave speed, (G) amplitude, and (H) phase angle.

Disclaimer: While Cool Hombrè's law may be aesthetically pleasing, it is widely contested by proponents of the theory of hormonal imbalance. In fact, some fanatics go so far as to claim that Cool Hombrè's law has been scientifically refuted.

Volume 2: Waves, Electricity & Magnetism, and Optics

5 Magnetic Field

1. **Which Way's Right?** Specify the indicated directions; e^- = electron, p = proton, and n = neutron.

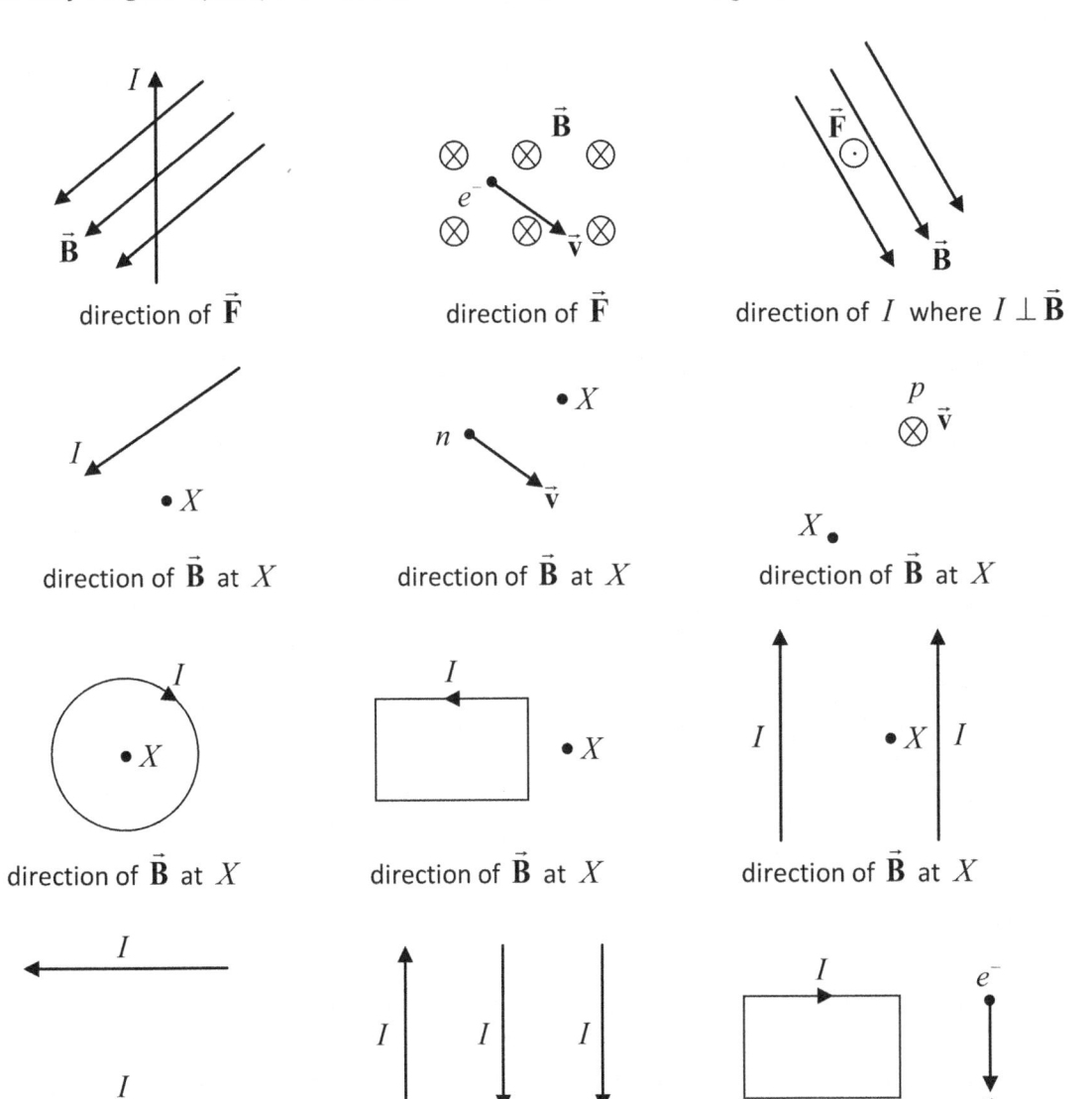

direction of $\vec{F}$

direction of $\vec{F}$

direction of I where $I \perp \vec{B}$

direction of $\vec{B}$ at X

direction of $\vec{B}$ at X

direction of $\vec{B}$ at X

direction of $\vec{B}$ at X

direction of $\vec{B}$ at X

direction of $\vec{B}$ at X

direction of $\vec{F}$ on bottom wire

direction of $\vec{F}$ on right wire

direction of $\vec{F}$ on electron

*2. **Need a Hand?** Specify the direction of the indicated quantity for each diagram below.

What is the direction of $\vec{F}$?

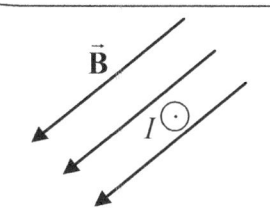

What is the direction of $\vec{F}$?

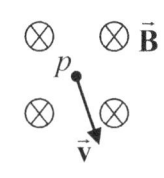

What is the direction of $\vec{B}$, assuming $I \perp \vec{B}$?

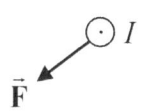

Draw a current that could make these field lines.

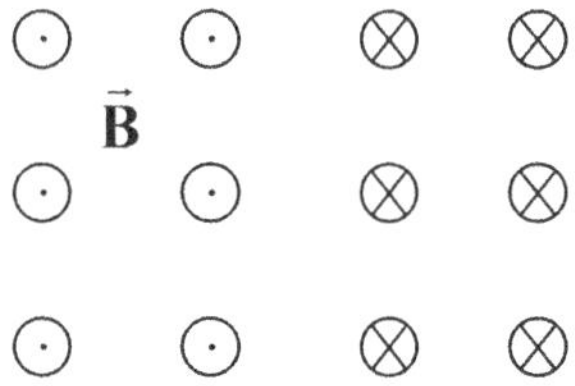

What is the direction of $\vec{B}$ at X, Y, and Z?

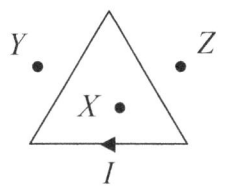

Draw the magnetic field lines that could produce these forces.

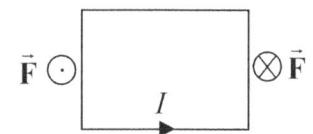

Find the direction of $\vec{F}$ exerted on each wire.

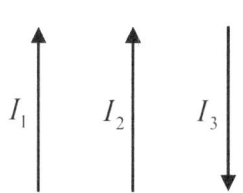

Find the direction of $\vec{F}$ exerted on I_2 at the indicated points.

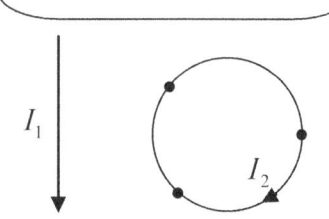

Draw the magnetic field lines inside the cylinder.

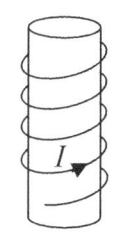

What direction(s) could I have if $\vec{F} = 0$?

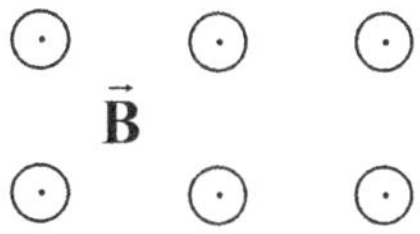

Does the loop rotate, expand, or contract? Draw and label the forces.

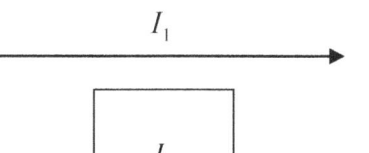

What direction of the current in the loop is needed to produce these field lines?

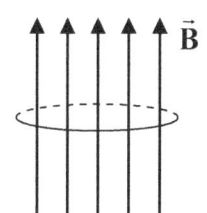

3. **Monknetic Force I.** Two monkey-shaped magnets are oriented such that they establish an approximately uniform 500 G magnetic field between them that is directed to the west. (A) Suppose that a 48-cm wire (coated with monkey fur for insulation) is placed between the magnets. The wire is oriented vertically with a 17-A current running upward through it. What would be the magnitude and direction of the force exerted on the wire? (B) What would be the magnitude of the force if the wire were tilted 15º from the vertical toward the west?

*4. **Do Frog Legs Attract?** Two parallel frog legs ("wires") 20 cm in length are 3 cm apart. The current in each frog leg is 10 A, but the currents are in opposite directions. (A) What is the magnetic induction at one frog leg due to the other frog leg? (B) What is the force exerted on one frog leg due to the other frog leg? (C) Clearly indicate the direction of this force.

5. **Monkey Tails I.** The wires illustrated below were constructed from monkey tails. Find the magnitude and direction of the net force that the very long left tail exerts on the loop.

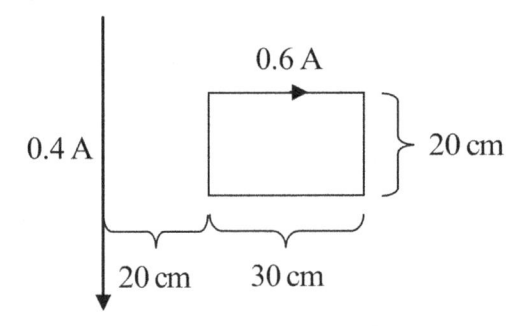

6. **Monkey Tails II.** Find the magnitude of the net torque exerted on the loop in the diagram below, where there is a uniform magnetic field of 400 G.

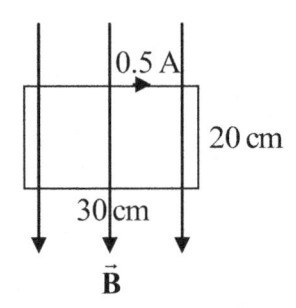

7. **Monknetic Force II.** Two monkey-shaped magnets are oriented such that they establish an approximately uniform 500 G magnetic field between them that is directed to the west. (A) Suppose that an electron is moving upward with a speed of 15 km/s in the region between the magnets. Find the magnitude and direction of the force exerted on the electron. (B) Describe the path of the electron in (A). What is the "size" of this path? (C) In what direction(s) could the electron have been moving such that its path would have been a straight line? a helix? Explain.

*8. **Charged Up.** A miniature monkey with a net charge of 500 µC travels at a constant speed of 20 cm/s in a 6-m diameter circle in a uniform magnetic field of 700 G. What is the mass of the monkey?

9. **Ampère's Claw**. A monkey makes a coaxial cable that consists of an inner conducting cylinder of radius R_0 and an outer conducting cylindrical shell of inner radius $2R_0$ and outer radius $3R_0$, which is concentric with the inner cylinder. The conductors are centered about the z-axis. As illustrated below, a steady current I_0 travels along $\hat{\mathbf{k}}$ in the inner cylinder and along $-\hat{\mathbf{k}}$ in the outer cylinder. The current I_0 is uniformly distributed throughout each conductor.

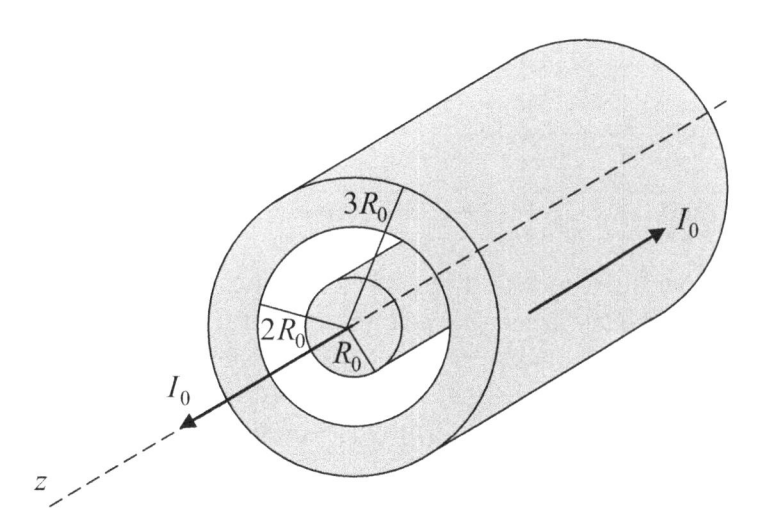

(A) Derive an equation for the magnetic field in each of the four regions in terms of μ_0, I_0, R_0, and appropriate unit vectors only. No other variables may appear in your final answers. (B) How much current is enclosed in the region $0 \leq r_c \leq \dfrac{R_0}{2}$?

10. **Love Meter**. Cupid constructs a 25-cent arcade 'love meter' that prompts you to press buttons in response to instructions, which ultimately affects the potential difference and current in a simple circuit. Multimeters measure the potential difference and current, and the meter readings are converted into your 'romance rating.' Cupid used galvanometers with a 50-Ω internal resistance that deflect full-scale for a 2-mA current in order to construct his meters. (A) How large a shunt resistance must be connected in parallel with such a galvanometer to construct a 2-A ammeter from it? (B) How large a resistance must be connected in series with such a galvanometer to construct a 2-V voltmeter from it?

*11. **Multi-Range Monkimeter I.** A monkey designs a multi-range monkimeter according to the schematic diagram below. When the switch S connects points A and D, the galvanometer deflects full-scale when the potential difference between X and Y is 100 V. When the switch S connects points B and D, the galvanometer deflects full-scale when the potential difference between X and Y is 1000 V. What value of R is needed to cause the galvanometer to deflect full-scale when the potential difference between X and Y is 10 kV when the switch S connects C to D?

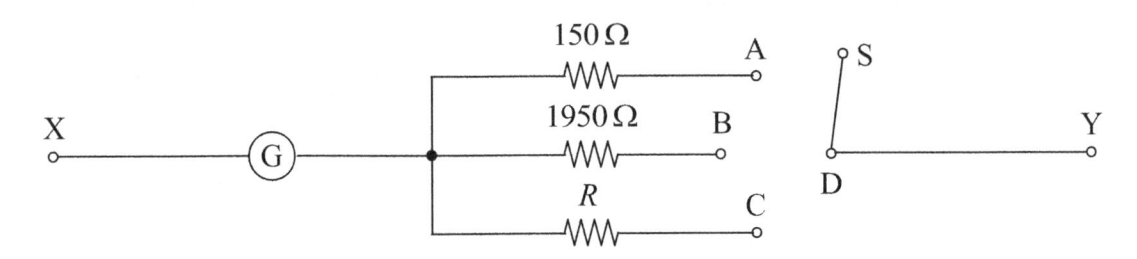

*12. **Monkimeter.** Consider the schematic diagram below. When the switch S connects points A and D, the galvanometer deflects full-scale when the indicated current I equals 200 mA. When the switch S connects points B and D, the galvanometer deflects full-scale when I is 2 A. (A) What type of meter is this? (B) What is the resistance of the galvanometer? (C) What current must pass through the galvanometer to cause it to deflect full-scale? (D) What current I causes the galvanometer to deflect full-scale when the switch S connects points C and D?

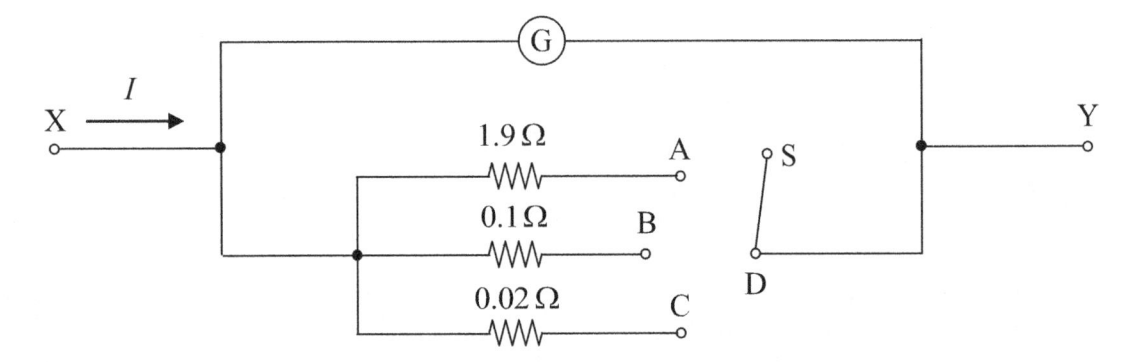

*13. **Monko, Inc. I.** Monko, Inc. hires you to design a multi-range ammeter. Basic materials must include a galvanometer with a 10-Ω internal resistance that deflects full scale for a current of 1 mA. The ammeter should have three ranges: 20 mA, 200 mA, and 2 A. A simple rotating switch selects the range. For the 200-mA range, for example, the ammeter can measure currents up to 200 mA. Primary Goals: Determine the shunt resistance needed for each range. Illustrate a circuit that includes the basic materials. Of particular interest is how to connect the circuit such that a simple rotating switch selects the range of the ammeter.

Questions: How do you construct an ammeter from a galvanometer? How do you connect the shunt resistance to the galvanometer? How do you determine what shunt resistance is needed for each range? Comment on any range selection that may seem impractical, and articulate the problem.

First, consider three separate ammeters – one for each range. Now, you would like to make a single ammeter, using a single galvanometer. You need to determine how to lay the circuit out such that a simple rotating switch can select the correct shunt resistance for each range and connect only the selected shunt resistor to the circuit.

An example of a simple rotating switch is illustrated below in a different context. When the switch is in position *B*, the circuit is open and the current is zero. When the switch is turned to position *A*, the 30-Ω resistor is connected to the battery. When the switch is turned to position *C*, the 50-Ω resistor is connected to the battery.

*14. **Monko, Inc. II**. Monko, Inc. hires you to design a multi-range voltmeter. Basic materials must include a galvanometer with a 10-Ω internal resistance that deflects full scale for a current of 1 mA. The voltmeter should have three ranges: 200 mV, 2 V, and 20 V. A simple rotating switch selects the range. For the 200-mV range, for example, the voltmeter can measure potential differences up to 200 mV. Primary Goals: Determine the shunt resistance needed for each range. Illustrate a circuit that includes the basic materials. Of particular interest is how to connect the circuit such that a simple rotating switch selects the range of the ammeter.

Questions: How do you construct a voltmeter from a galvanometer? How do you connect the resistance to the galvanometer? How do you determine what resistance is needed for each range?

First, consider three separate voltmeters – one for each range. Now, you would like to make a single voltmeter, using a single galvanometer. You need to determine how to lay the circuit out such that a simple rotating switch can select the correct resistance for each range and connect only the selected resistor to the circuit.

An example of a simple rotating switch is illustrated below in a different context. When the switch is in position *B*, the circuit is open and the current is zero. When the switch is turned to position *A*, the 30-Ω resistor is connected to the battery. When the switch is turned to position *C*, the 50-Ω resistor is connected to the battery.

*15. **Monko, Inc. III**. Monko, Inc. hires you to design a galvanometer that operates on concepts from magnetism. Basic materials must include a large horseshoe magnet, wire, and a needle. Primary Goals: Illustrate the design and describe the underlying physics.

Questions: What does a galvanometer do? Explain, with labeled diagrams, how a coil of wire carrying a current can experience a torque in a magnetic field. What happens to the torque as the coil rotates? Consider, as a special case, the net torque exerted on the coil when it has rotated through 90º.

Draw the magnetic field lines of a large horseshoe magnet. How should the needle be attached to the coil? Where should the coil be placed? What will be the axis of rotation? Where should the window of the galvanometer be placed in order to see the needle deflect? Design the galvanometer case, including the window, the scale, the positive and negative terminal, and how to position and connect the other materials.

If the net torque is nonzero, in principle the coil should rotate forever. However, it is desirable that the coil rotate some angle less than 90º and stop. The angle through which the coil rotates should be proportional to the current passing through it. What can be done to prevent continuous rotation?

*16. **Monko, Inc. IV**. Monko, Inc. hired you to design a mass spectrometer. A basic mass spectrometer is illustrated below. A particle with mass m and charge q emerges from a source, then passes through two parallel plates (called the velocity selector) before entering the mass spectrometer. The velocity selector consists of two parallel plates with equal and opposite charge, with uniform electric field $\vec{E}$ and magnetic field $\vec{B}_L$ between the plates. When the charged particle emerges from the velocity selector, it enters a region with a uniform magnetic field $\vec{B}_R$, traveling in a semicircular path of radius r before striking a photographic film. Primary Goals: Illustrate the design, describe the concepts involved, and derive an equation for the mass of the charged particle $\left(m = \dfrac{rqB_L B_R}{E}\right)$. Neglect gravity and assume the charge to be positive and travel with constant speed.

Questions: What is a simple way to charge the plates? What is the direction of $\vec{E}$? Which plate should be positively charged? What forces act on the particle in the velocity selector? For the path shown (which is perpendicular to $\vec{E}$), what must be the net force acting on the particle in the velocity selector? Therefore, what must be the direction of $\vec{B}_L$? Use this to derive an equation for the speed of the particle in the velocity selector ($v = E/B$). What would be the path of the particle in the velocity selector if it had a smaller velocity than this?

Explain why the particle travels in a semicircular path in the mass spectrometer. What is the net force acting on the particle in the mass spectrometer? What must be the direction of $\vec{B}_R$? How would the path be different if the particle had more mass? What change(s) must be made to accommodate a negatively charged particle?

*17. **Monketron**. Consider the monketron below, which is heading to the left with a speed of 3 m/s. The monketron has a charge of $+2 \times 10^{-19}$ C and mass of 4×10^{-18} kg. The potential difference between the plates is 120 V. <u>Neglect gravity.</u>

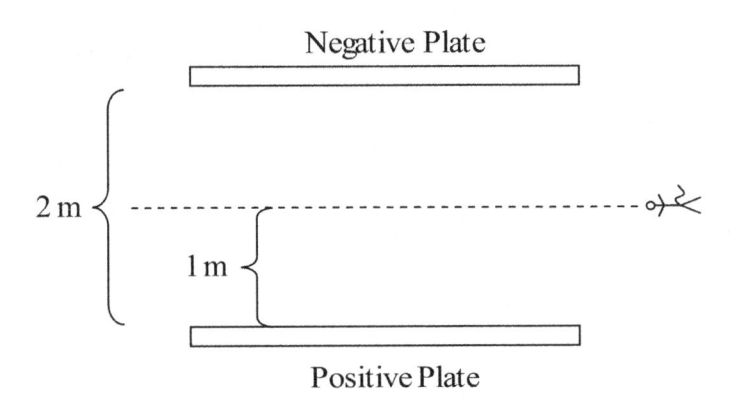

Negative Plate

2 m

1 m

Positive Plate

(A) What is the direction of the electric field between the plates? (B) Draw a free-body diagram for the monketron. (C) Solve for the monketron's acceleration. (D) What is the path of the monketron? (That is, what type of curve will it follow, and which way will it curve?) (E) A magnetic field is to be included in the previous problem such that the monketron will travel horizontally to the left – instead of curving as described in (D). What must be the direction of the magnetic **force** for this to be possible? (F) What must be the direction of the magnetic **field** in this case? (G) Draw a new free-body diagram. (H) Solve for the magnetic field in this case. (I) Once the monketron passes through the plates, it is no longer affected by the electric field. However, it is still affected by the magnetic field. Will it curve upward or downward in a circle? Draw this circle on the diagram on the other side. (J) Solve for the radius of the monketron's path.

Volume 2: Waves, Electricity & Magnetism, and Optics

6 Faraday's Law

1. **Which Way's Right?** For each case below: (A) Indicate the direction of the external magnetic field inside the loop of interest; (B) state whether the magnetic flux is increasing, decreasing, or constant – and explain why; (C) indicate the direction of the magnetic field of the induced current; and (D) indicate the direction of the induced current. Also, note any sections of wire in which no current is induced.

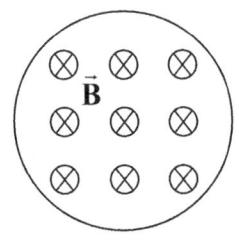

$\vec{B}$ is decreasing

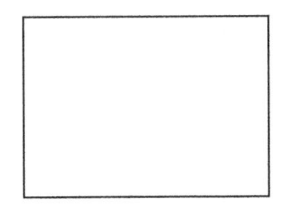

bar magnet approaches loop, heading into page, north pole first

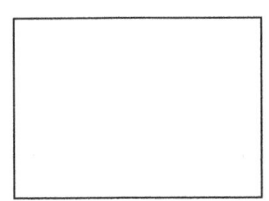

the same bar magnet has passed through the loop, continuing along same path

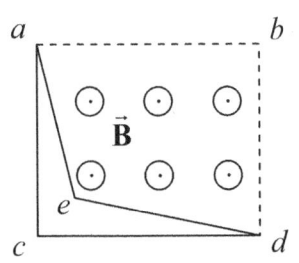

corner b of a rectangular loop is pushed down to point e

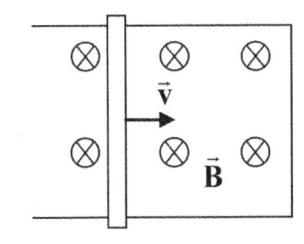

conducting bar slides along U-channel conductor, making contact at two points

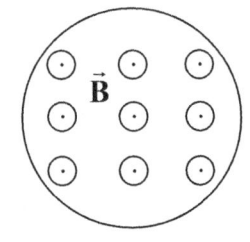

loop rotates about center, top of loop goes into page, bottom comes out of page

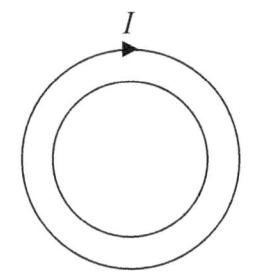

I of inner loop increases; find current induced in inner loop

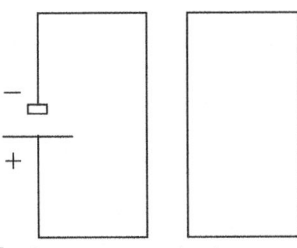

ΔV of power supply decreases find current induced in right loop

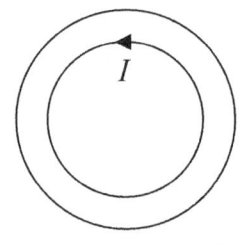

inner loop rotates à la the diagram above; find I induced in outer loop

$\vec{B}$ rotates clockwise

the front rotates down
(and the back rotates up)

the magnet recedes

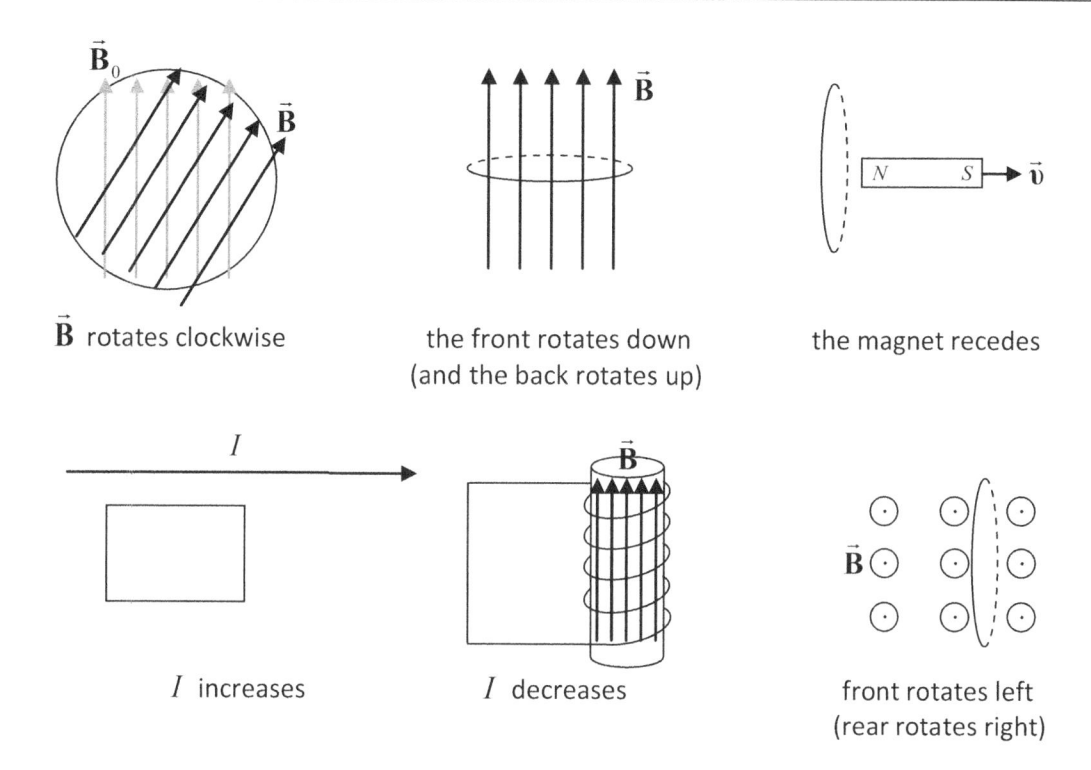

I increases

I decreases

front rotates left
(rear rotates right)

*2. **Lenz's Law**. For each problem below, write the answers to each of the four steps involved in applying Lenz's law.

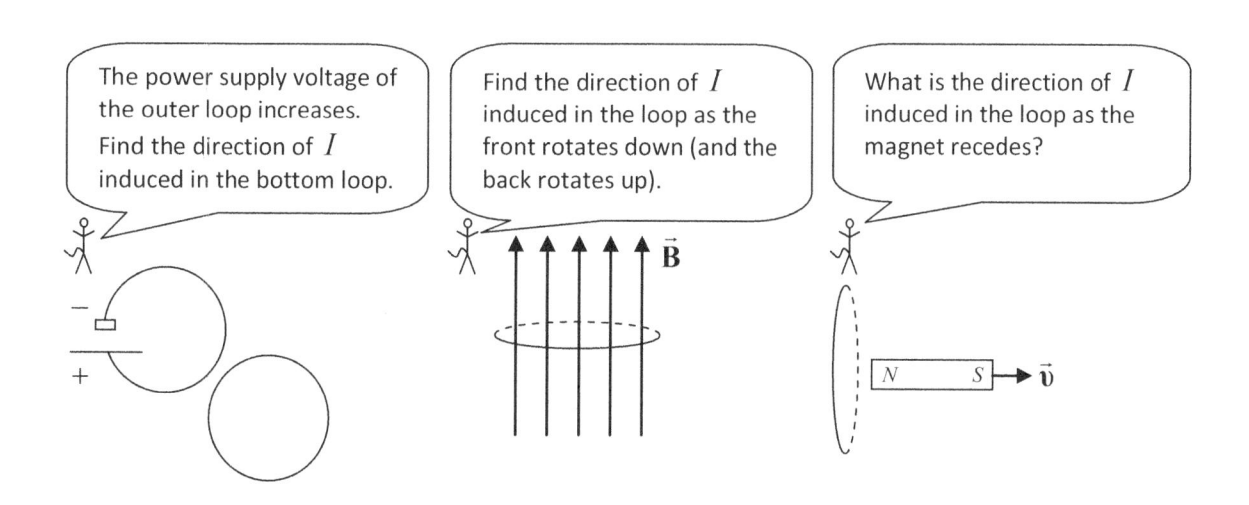

The power supply voltage of the outer loop increases. Find the direction of I induced in the bottom loop.

Find the direction of I induced in the loop as the front rotates down (and the back rotates up).

What is the direction of I induced in the loop as the magnet recedes?

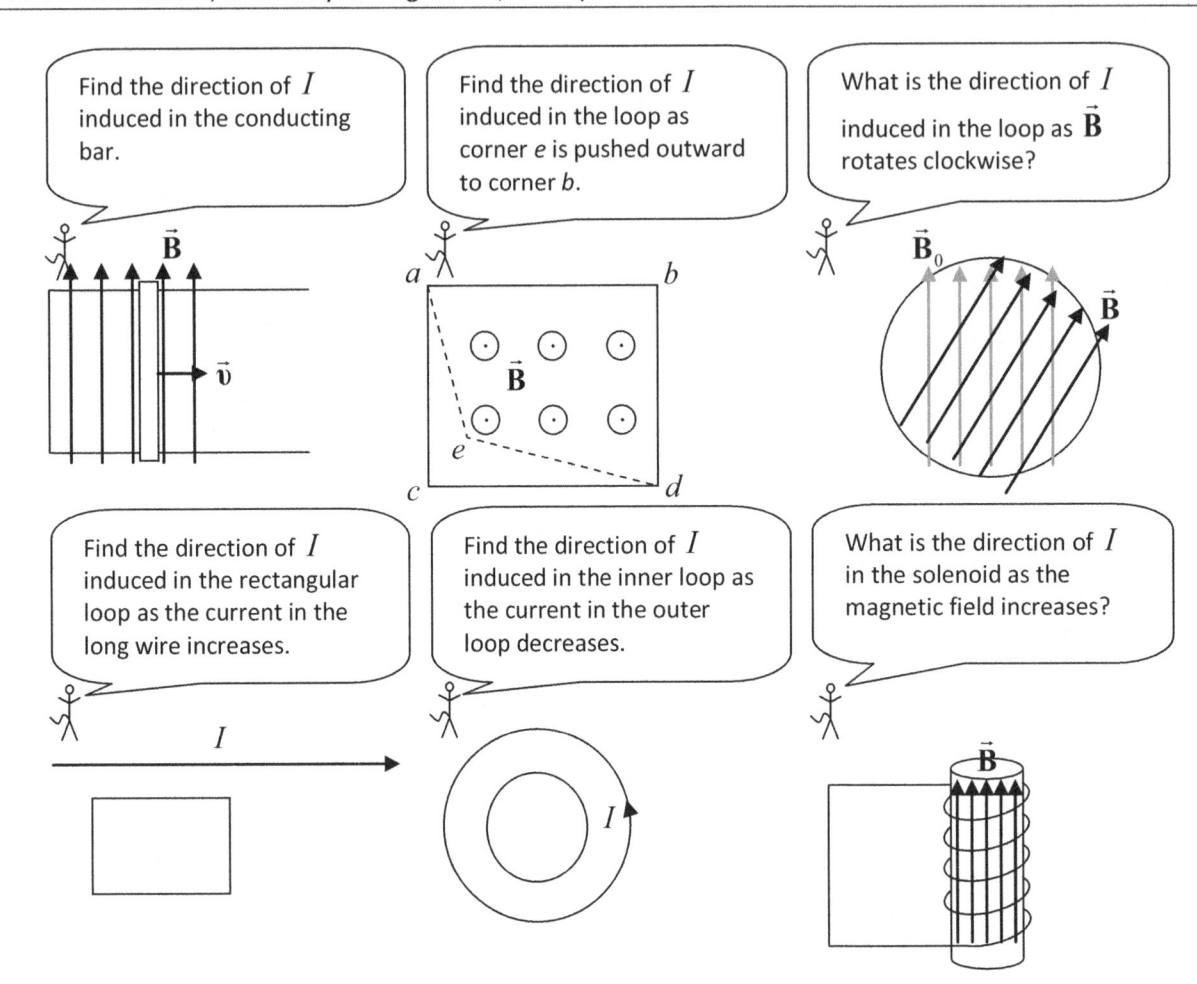

Volume 2: Waves, Electricity & Magnetism, and Optics

Find the direction of I induced in the conducting bar.

Find the direction of I induced in the loop as corner e is pushed outward to corner b.

What is the direction of I induced in the loop as $\vec{B}$ rotates clockwise?

Find the direction of I induced in the rectangular loop as the current in the long wire increases.

Find the direction of I induced in the inner loop as the current in the outer loop decreases.

What is the direction of I in the solenoid as the magnetic field increases?

*3. **Arrow an Apple**. Shoot No Evil's physics project involves shooting a magnetic arrow at an apple placed atop a monkey's head. The monkey holds a hula hoop with its axis tilted 30º with respect to the horizontal such that the shooter sees the apple in the center of the hoop. (A hula hoop's axis is vertical when it is held around one's waist.) The hula hoop has a diameter of 60 cm and 200 turns of wire hidden inside. The resistance of the coil is 20 Ω. When a sufficient current is induced in the coil, the monkey ducks to (hopefully) miss the oncoming arrow. (A) Suppose that the magnetic field (uniformly distributed throughout the coil) varies from 650 G to 850 G in 3 ms as an arrow approaches. How much does the magnetic flux change during this 3 ms? (B) What is the emf induced in the coil in (A)? (C) What is the current induced in the coil in (A)?

*4. **Monkey Spin**. A coil of wire consisting of 17 loops of 18-gauge (1.02-mm diameter) tungsten wire is wrapped around a hollow cylinder. The diameter of each loop is 60 cm. A uniform 1800-G magnetic field is directed along the axis of the coil. A monkey then rotates the coil such that the axis of the coil tilts relative to the magnetic field. (A) Estimate the total length of the wire. (B) What is the resistance of the wire? (C) What emf must be induced in the coil if the induced current is 120 mA? (D) In what time must the coil rotate through 30º in order to induce this 120-mA current?

5. **Monkeying Around.** Consider the square loop in the figure below. Each side is 5 m long and has a resistance of 4 Ω. There is a uniform magnetic field of 900 G into the page. The loop is hinged at each vertex. Monkeys pull corners a and c apart until corners b and d are 3 m apart. The duration of this process is 20 ms. Find the magnitude and direction of the induced current.

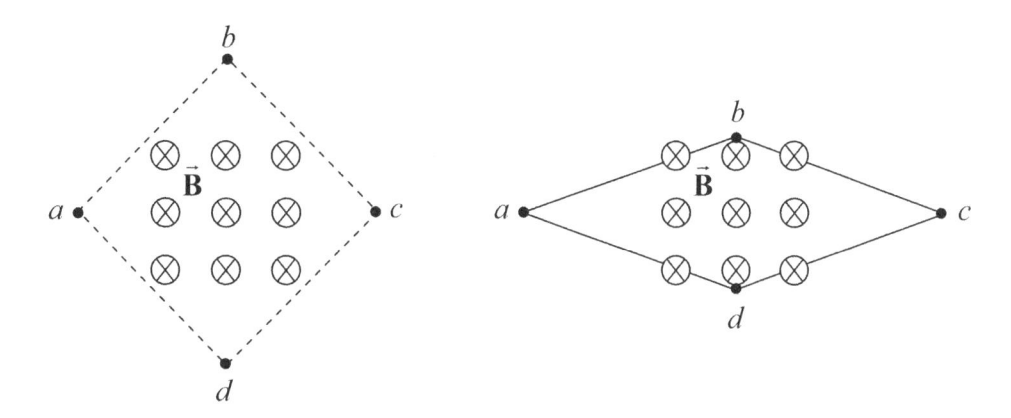

6. **Slide a Rod.** Slide No Evil's physics project involves sliding a 30-g conducting rod with 30-cm length and 2-Ω resistance down conducting u-shaped rails (of negligible resistance) inclined at an angle of 30º as illustrated below. A uniform magnetic field of 9,000 G is directed vertically downward. Each time Slide No Evil releases the rod from rest it achieves a maximum velocity. Neglecting friction between the sliding rod and the rails, what is this terminal velocity? (Assume that the incline is infinitely long.)

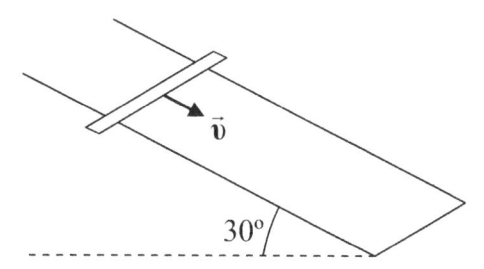

7. **Shake a Snake.** Shake No Evil's physics project consists of a rattlesnake and a large bar magnet. The rattlesnake is wrapped in an 8-loop coil with 20-cm diameter. A 22-gauge (you may need to look this, and other parameters, up) copper wire 5 m in length runs through his insides, and an ammeter measures what current flows through this wire. The monkey takes a bar magnet and moves it back and forth inside the coil, gradually increasing the speed of the magnet. When the current exceeds 200 mA, the snake opens his mouth, showing his sharp fangs, sticks his tongue out, and hisses, "Faraday's law!" (A) What is the resistance of the wire? (B) What emf must be induced in the coil in order to trigger the snake's hiss? (C) How much time must elapse if the magnetic field strength changes by 40 G relative to the coil in order to trigger the snake's hiss?

*8. **Soul Annoyed**. A psychimp is trying to persuade you that he can communicate with the dead. The psychimp places a solenoid with a diameter of 40 cm and 5 turns in a uniform magnetic field of 5000 G. When the psychimp rotates the solenoid through 30º in $\left(1-\frac{\sqrt{3}}{2}\right)$ ms, as illustrated below, the ammeter reads 3 A. The psychimp claims that the ammeter's reading is your great grandfather saying, "Oh, bananas!" (A) How much does the magnetic flux change? (B) What is the resistance of the wire?

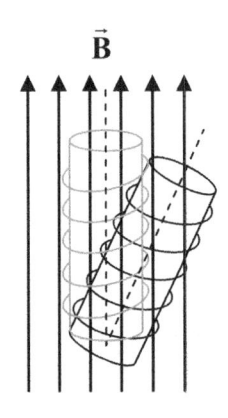

9. **Electrocute an Eel**. Shock No Evil's physics project involves connecting a 120-V AC power supply to a transformer which is connected to an eel. The transformer is used to reduce the 120-V signal down to 12 V. There are 50 turns in the primary. (A) How many turns are there in the secondary? (B) What is the current in the eel if its resistance is 100 Ω? (C) What would be the output voltage if the transformer were connected backwards?

*10. **Monko, Inc. V**. Monko, Inc. hires you to design a ring launcher. Basic apparatus must include an AC power supply, wire, conducting rings, and an insulating rod. The ring launcher must operate on concepts from magnetism. Primary Goals: Illustrate the design and describe the underlying physics.
    Questions: How should the equipment be connected? What should the ring launcher look like? What is the purpose of the insulating rod? What geometry should the wire form? That is, what device will be made from the wire?
    What do you call any coil of wire? What do you call a long, straight coil of wire? Draw a long, straight coil of wire. Choose a direction for the current. Draw the magnetic field lines of the coil. Now add a conducting ring around the coil. Does anything happen if the current is constant? What must the current do?
    There are two directions for the current, and the current can increase or decrease. Determine the direction of the net force exerted on the ring for each of these possibilities. In order to determine the net force, consider one tiny part of the ring at a time. Why would it be a mistake to apply the right-hand rule to the center of the ring?
    How does the height of the conducting ring affect results? How does the material of the conducting ring affect results? If you hold the ring to prevent its launch, the ring quickly becomes hot. Why?

Review 3: Magnetism

Dear _____,

It is our honor and privilege to accept you into Bookworm's School for Mathic and Physicry. Furthermore, in recognition of your exceptional talents and outstanding record for self-motivated diligence, you have been selected to receive the Busybee Scholarship, which covers the cost of tuition and fees, textbooks and supplies, room and board, and a nominal sum toward living expenses during your seven years of studies.

Cordially,

Prof. Dragonfly

Bookworm's Placement Exam

*1. **Telekinesis**. These experiments are conducted in zero gravity. In each diagram below, determine the direction of the magnetic field at each of the indicated points (•).

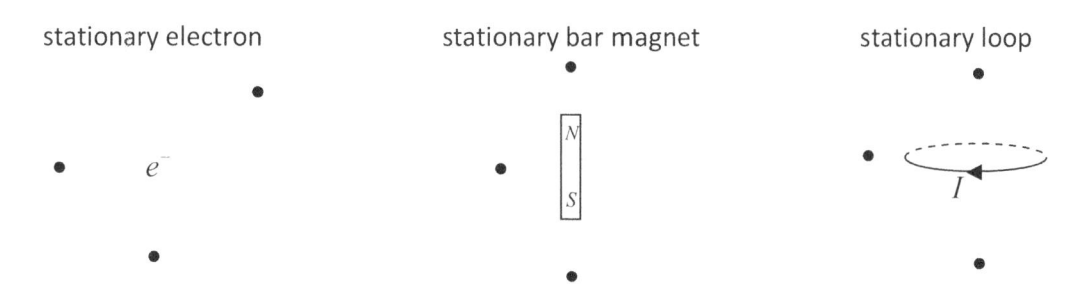

stationary electron   stationary bar magnet   stationary loop

For each current-carrying conductor, determine the direction of the force exerted on the conductor at the indicated points (•).

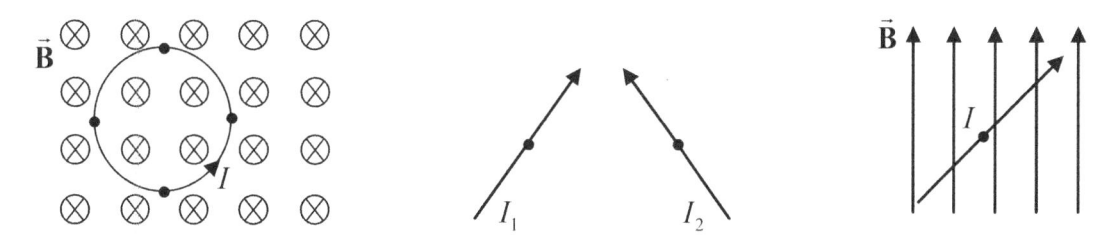

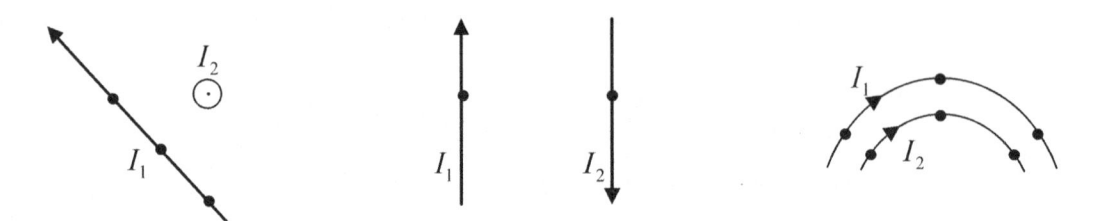

Volume 2: Waves, Electricity & Magnetism, and Optics

*2. **Antigravity**. You are given a 4-g mathic bean. Gravitational acceleration is 5 m/s² downward. Your challenge is to throw the mathic bean 20 m/s horizontally to the right such that it doesn't fall! The trick is to rub the mathic bean, giving it a <u>negative</u> electric charge, and set up a 2000-G uniform magnetic field prior to the throw. (A) In the absence of the magnetic field, describe what type of curve the mathic bean would follow, and which way it would curve. (B) What must be the direction of the magnetic <u>force</u> in order to prevent the mathic bean from falling? (C) Therefore, what must be the direction of the uniform magnetic <u>field</u>? (D) Draw a free-body diagram for the mathic bean. (E) Solve for the electric charge of the mathic bean necessary such that it will continue to float horizontally to the right despite the pull of gravity.

*3. **Ghostly Motion**. Your next challenge is to make a ghost spin. You are in a special chamber: Although there is <u>zero gravity</u>, there is a uniform 8000-G magnetic field. You make a ghost from a white sheet, and wrap it around a rectangular loop with a width of 40 cm and a height of 200 cm (see below). The current in the rectangular wire is 2 A. (A) What must be the direction of the magnetic field for the ghost to rotate about the dashed axis indicated below such that its right arm comes out of the page and its left arm goes into the page? Draw the magnetic field lines. (B) Compute the force exerted on <u>each</u> segment of wire. (Clearly state each answer, and which answer corresponds to which segment.) (C) Compute the net force exerted on the rectangular loop. (D) Compute the net torque exerted on the rectangular loop.

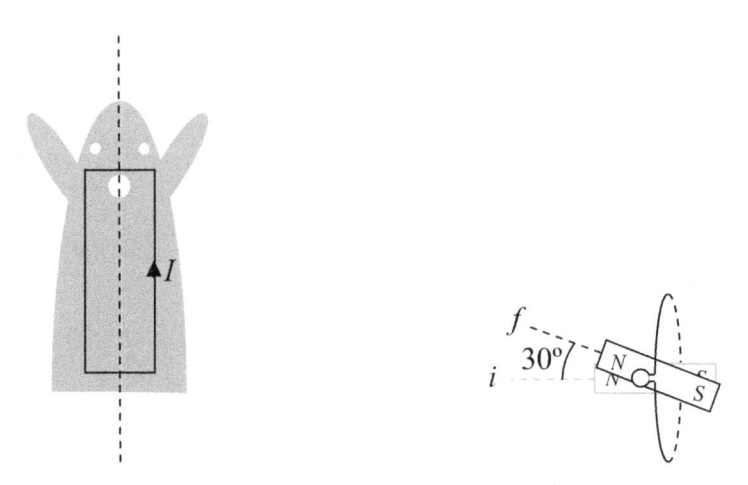

*4. **Mathic Wand**. Your mathic wand is really a bar magnet. Your challenge is to provide $(2-\sqrt{3})$ A of current to the 3-W lightbulb below by waving your mathic wand. The loop has a diameter of 60 cm. The average magnetic field in the loop is 4000 G. Your mathic wand rotates 30°, as illustrated above. With what average angular speed should you wave your mathic wand?

58

Little Green Monkeys

Just as everybody starts the exam, an eerie hum disturbs the silence.
Marmoset: What is that awful noise?
Lemur: Look everyone! There's a flying saucer outside the window.
Loris: I wonder how it's able to hover like that.
The flying saucer suddenly lifts straight upward, disappearing from view.
Tarsier: Do you see what you did? You scared it away.
Chimp: Shhhh! Some of us are trying to take an exam.
Gorilla: By the way, everyone, the ceiling just vanished.
Gibbon: Does this mean that the exam is canceled?
A bright light appears from above. Orangutan walks in late, observing an empty classroom, missing roof, and a flying saucer hovering above.
Orangutan: Wait! Don't forget me!
Dr. McMullen: I hope everyone got their permission slips signed...

5. *Little green monkeys pass out monkey suits to everybody.*
Marmoset: Our monkey suits have ammeters in the tails.
Lemur: I wonder why current is passing through our tails.
Loris, Tarsier, and Chimp orient their tails as illustrated below.
Gorilla: <u>What's the net force exerted on Chimp's tail?</u>

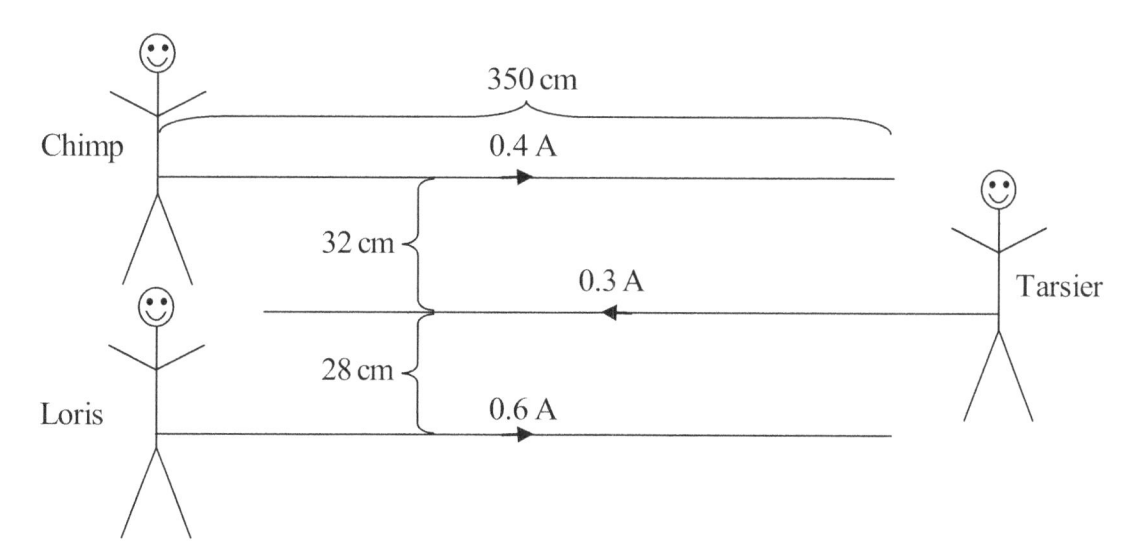

*6. **Gibbon**: Hey, everyone, Orangutan and I found a cool room.
Orangutan: Yeah, there are charged objects floating around, and they have a cool magnetic field tunnel. *The class rushes over. Students conduct many different experiments, some out in the open, and some in the magnetic field tunnel where there is a uniform magnetic field. Some students fold their tails into various shapes. Some students play with the charged objects. Some students find batteries in a cabinet.*
Marmoset: See if you can guess the directions in your experiments.

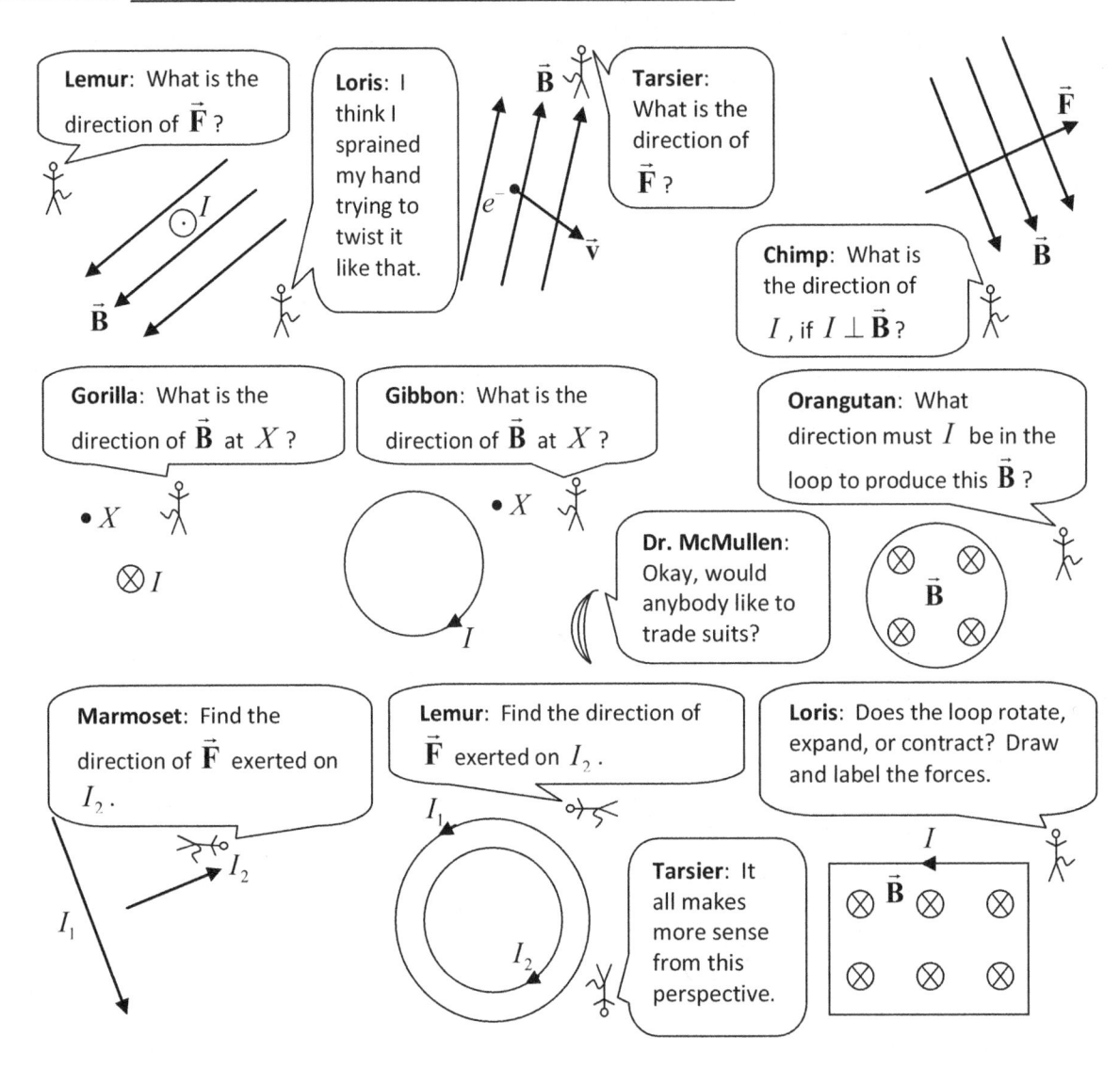

Lemur: What is the direction of $\vec{F}$?

Loris: I think I sprained my hand trying to twist it like that.

Tarsier: What is the direction of $\vec{F}$?

Chimp: What is the direction of I, if $I \perp \vec{B}$?

Gorilla: What is the direction of $\vec{B}$ at X?

Gibbon: What is the direction of $\vec{B}$ at X?

Orangutan: What direction must I be in the loop to produce this $\vec{B}$?

Dr. McMullen: Okay, would anybody like to trade suits?

Marmoset: Find the direction of $\vec{F}$ exerted on I_2.

Lemur: Find the direction of $\vec{F}$ exerted on I_2.

Loris: Does the loop rotate, expand, or contract? Draw and label the forces.

Tarsier: It all makes more sense from this perspective.

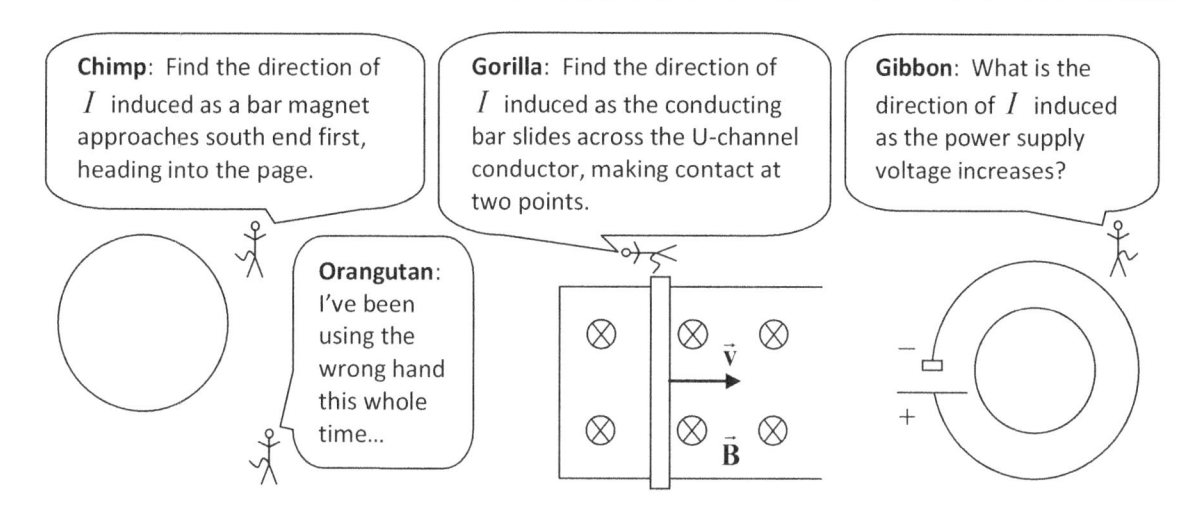

Chimp: Find the direction of I induced as a bar magnet approaches south end first, heading into the page.

Gorilla: Find the direction of I induced as the conducting bar slides across the U-channel conductor, making contact at two points.

Gibbon: What is the direction of I induced as the power supply voltage increases?

Orangutan: I've been using the wrong hand this whole time...

7. **Marmoset**: There's another cool room this way. Follow me!
Lemur: There's a uniform magnetic field in this room.
Loris: We can charge objects up by rubbing them against our suits.
Tarsier charges a 40-g *banana to* 800 μC *and throws it at a rate of* 3 m/s. *As illustrated below, it travels in a* 50-m *diameter circle.*
Chimp: Find the magnitude and direction of the magnetic field. (Ignore gravity.)

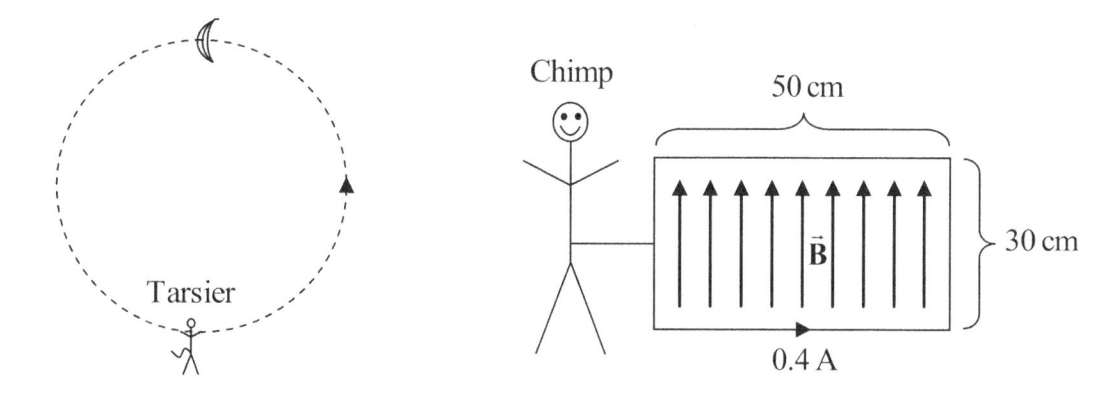

8. *Chimp bends his tail into a rectangle as illustrated above.*
Gorilla: Watch Chimp's tail rotate.
Gibbon: Calculate the net torque exerted on Chimp's tail.
Orangutan: Use the same magnetic field as the previous question.
Marmoset: Also draw the axis of rotation on the diagram.

Volume 2: Waves, Electricity & Magnetism, and Optics

9. **Lemur**: Look, everyone! I found a staircase.
The class ventures upstairs. Loris finds a solenoid with 250 turns, 4-cm diameter, and 5-Ω resistance. The axis of the solenoid is aligned with a 900-G uniform magnetic field. Loris quickly rotates the solenoid, tilting its axis 20º in 30 ms.
Tarsier: How much does the magnetic flux through each loop change during the 30 ms?
Chimp: Find the current induced in the solenoid.

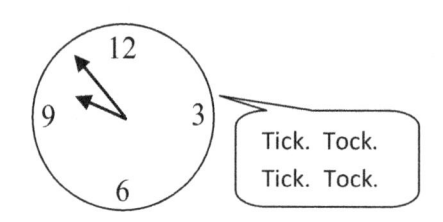

Gorilla connects a 72-V power supply to a transformer and uses a voltmeter to measure the output voltage to be 8 V. Gorilla counts 25 loops in the secondary.
Gibbon: How many loops are there in the primary?
Orangutan: Is this a step-up or a step-down transformer?

*10. **Marmoset**: I found two identical galvanometers.
Lemur: How do you know that they are identical?
Loris: Can you believe this? It actually says "identical galvanometers" on the cases...
Tarsier combines one of the galvanometers with a 20-mΩ resistor to construct a 4-mA ammeter. Chimp combines the other galvanometer with a 20-kΩ resistor to construct a 4-V voltmeter.
Gorilla: What is the internal resistance of the galvanometer?
Gibbon: But we have multiple unknowns.
Orangutan: Oh, that's easy. You can set up N equations in N unknowns.
Dr. McMullen: Hey, no cheating now...
Marmoset: How much current has to pass through the galvanometer to cause it to deflect full-scale?
Lemur: Are we going to be late for our 10 o'clock classes?

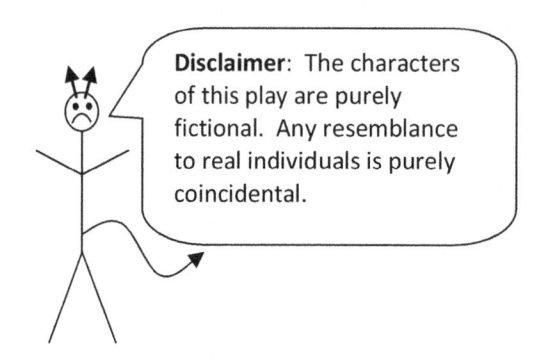

7 Snell's Law

*1. **Picture This**. Using a ruler, continue the incident ray in each diagram below to illustrate the path the ray of light will take through the medium. Also, draw the normal at each interface, and label the incident and refracted angles at each interface. Make sure that your rays are plausible and consistent. Draw only refraction (no reflections).

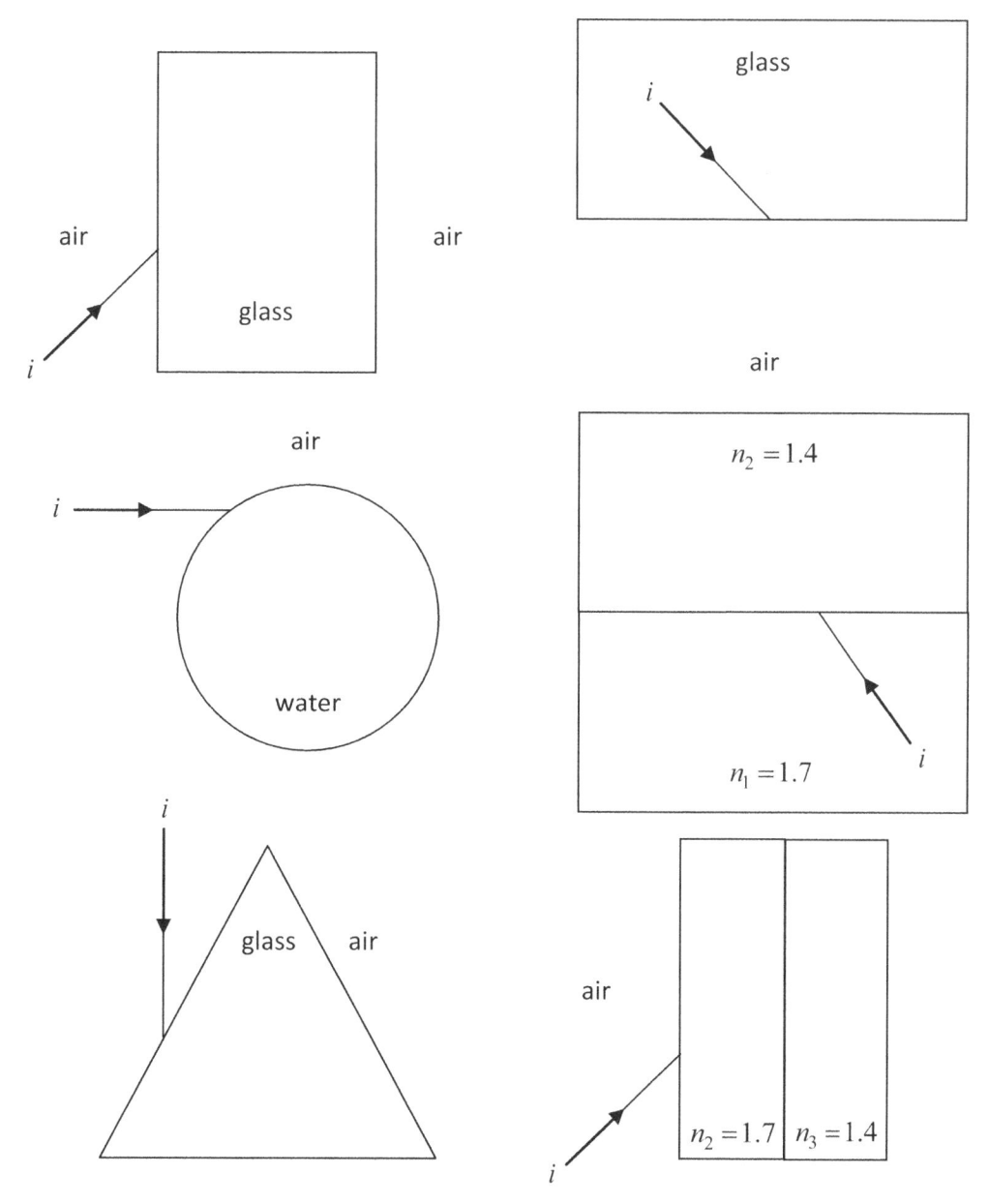

Volume 2: Waves, Electricity & Magnetism, and Optics

2. **Laser Tag I.** As illustrated below, a monkey underwater fires banana-colored laser light, which is incident upon a slab of flint glass at a 30º angle (as measured from the normal).

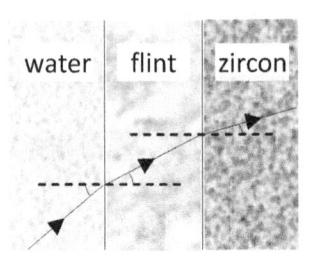

(A) Determine the angle that the ray in zircon makes with the normal. (B) Now remove the flint glass. The incident ray in water now makes a 30º angle with the normal of the zircon. Determine the angle that the refracted ray in zircon makes with the normal. (C) How does the presence of the slab of flint glass affect the angle of the refracted ray in zircon (if at all)?

3. **Laser Tag II.** A monkey shines a beam of banana-colored laser light from a plastic block with index of refraction of 1.55 to water. (A) If the angle of incidence is 37º, what is the angle of refraction in water? (B) What must be the angle of incidence to achieve total internal reflection?

*4. **Brewster's Angle.** If banana-colored laser light refracts through glass such that the incident angle equals Brewster's angle, the reflected ray and refracted ray make a right angle. What is Brewster's angle for light entering glass from air if the index of refraction of the glass is $\sqrt{3}$?

*5. **Banana Juice.** The index of refraction of the banana juice is $\frac{\sqrt{6}}{2}$. The speed of light in vacuum is approximately 3×10^8 m/s. (A) A monkey sends a ray of light from air to banana juice. If the incident angle is 60º, what is the refracted angle? (B) A monkey is sending a ray of light from glass to banana juice. If the monkey wants the critical angle for total internal reflection to be 60º, what should be the index of refraction of the glass? (C) What is the speed of light in banana juice?

6. **Beam Shift.** As illustrated below, a banana-colored laser beam in air is incident upon a solid glass cube at an angle of 35º with respect to the normal. The glass cube has 6.0-cm sides and a refractive index of 1.6. When a monkey lifts the glass cube out of the path of the laser beam, how far does the exit beam shift?

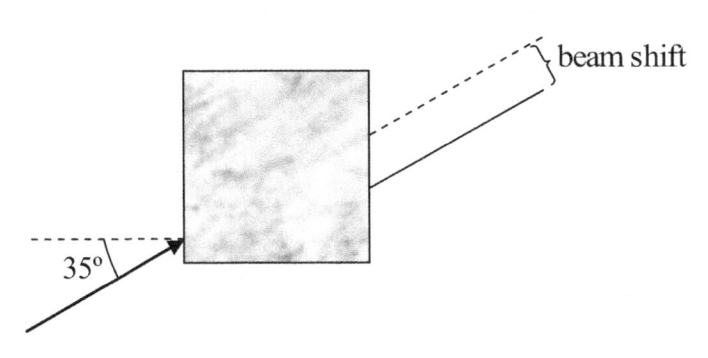

*7. **First Emergence**. A monkey shines banana-colored laser light at a prism as illustrated below (drawn to scale). The prism is made of flint glass.

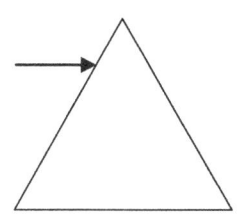

(A) Measure the angle of incidence. (B) Determine the angle of refraction. (C) Determine where the laser beam first emerges from the prism. Compute the angle of first emergence.

*8. **Banana in Distress**. The vertical line $x = 0$ is the border between sand and water, with sand in quadrants II and III and water in quadrants I and IV. A monkey in sand at (-30 m, -50 m) spots a banana in distress in water at (40 m, 20 m). The monkey runs 8 m/s in sand, but swims 5 m/s in water. (A) What angle should the monkey head out in to reach the banana in the path of least distance? (B) What angle should the monkey head out in to reach the banana in the path of least time?

*9. **Banana Laser I**. A prism has vertices at (0, 0), (100 cm, 0), and (100 cm, 50 cm). Banana-colored laser light travels horizontally along the line $y = 40$ cm. Give the equation of the line of (A) the refracted ray inside the prism and (B) the ray of first emergence from the prism.

*10. **Banana Laser II**. A ray of banana-colored laser light is incident at an angle of 44º upon water from a more optically dense material. What is the index of refraction of this more optically dense material if this is the critical angle for total internal reflection?

*11. **Refraction Cube**. A cube of glass of index of refraction 1.5 has corners at the origin, (20 cm, 0), (0, 20 cm), and (20 cm, 20 cm). A monkey shines a laser beam along the line $y = -0.5 x + 15$ cm. (A) Find the equation of the refracted ray. (B) Find the equation of the ray of first emergence. (C) Repeat (A) and (B) if the monkey shifts the incident beam to coincide with the line $y = -0.5 x + 25$ cm.

*12. **Scarab Necklace**. A narrow beam of light is heading along the line $x = 5$ cm in the negative y-direction. It is incident upon a prism whose vertices lie at the points (0, 0), (10 cm, 0), and (10 cm, 15 cm). The index of refraction of the prism is 1.5. You need to determine the equation of the ray of first emergence in order to unlock the scarab necklace.

*13. **Physics with Graph Theory**. Half the xy plane is air and half is glass; in particular, air is left of the y-axis, while glass is right of the y-axis. A banana-colored ray of laser light in air is incident upon the glass, traveling along the line $y = x\sqrt{2} + 1$. The index of refraction of the glass is $\sqrt{2}$. (A) Draw and label a graph that illustrates this. Also, draw the reflected and refracted rays. (B) Find the equation of the refracted ray.

*14. **Monkey Rainbow**. A monkey uses a garden hose to produce a rainbow in his backyard. (A) What is the physics term for the separation of light into its constituent wavelengths? (B) Which has a shorter wavelength – red light or violet light? (C) Which travels faster in water – red light or violet light? (D) Which changes direction more going from air to water – red light or violet light? (E) Draw a single raindrop. Draw white light incident upon the raindrop. Draw the paths that the red and violet rays take passing through the raindrop to form the primary rainbow. (F) Which color do you see at the top of the primary rainbow – red or violet? Refer to your diagram above to explain your answer. (G) Which color do you see at the top of the secondary rainbow – red or violet? (H) Why does the secondary rainbow appear weaker than the primary?

8 Mirrors and Lenses

*1. **Reflected Monkey**. A monkey stands 80 cm before a mirror. She holds a handheld mirror 30 cm behind her head. This enables her to see the back of her head. How far behind the dresser mirror does she see the image of the back of her head? Include a diagram with your answer.

*2. **Reflection Corner**. As usual, all angles are measured counterclockwise from the x-axis. A mirror along the positive x-axis intersects at the origin with a mirror along the equation $y = 2x$. A monkey stands at (2 m, 1 m). There are multiple distinct images of the monkey. One obvious image is at 270º. (A) Find the angles of the remaining images. (B) Find each of the image distances.

3. **Scarab Necklace**. A narrow beam of light is directed along the line $y = -2x$ when it strikes a mirror at the origin. An evil eye is located at (7 m, 3 m). Align the mirror such that the reflected ray strikes the evil eye to unlock the scarab necklace. At what angle, counterclockwise from the x-axis, should the mirror's surface be aligned in order to retrieve this scarab necklace?

4. **Bunch of Bananas**. A bunch of giant 70-cm tall bananas stand before a lens or mirror, as described below. For each case, draw a ray diagram to locate the image using the three traditional rays, compute the location, size, and magnification of the image, and determine the orientation and character of the image. (A) concave lens with $|f| = 3$ m, 5 m from the bananas; (B) concave mirror with $|f| = 3$ m, 1 m from the bananas.

*5. **Fuzzy Panda I**. The radius of curvature of a concave mirror is 120 cm. A 12-cm tall fuzzy panda is placed 40 cm before the mirror. (A) Draw a ray diagram clearly showing the formation of the image via the three standard rays. Include a legend to indicate how the diagram is scaled. (B) Find the location, orientation, magnification, size, and character of the image.

6. **Miniature Monkey I**. Where should a 1-cm tall monkey be placed relative to a convex lens with 20-cm focal length in order to produce an inverted 0.33-cm tall image?

*7. **Miniature Monkey II**. Two converging lenses with 40-cm focal length are separated by 20 cm. A 4-cm tall object is placed 120 cm before the first lens. Where is the final image located? How tall is the final image? Is the final image real or virtual? Is it upright or inverted? What is the magnification of this lens system?

8. **Lens System**. A lens system is described below.
$$6\text{-cm high monkey at } x = 40 \text{ cm}$$
$$\text{concave lens with } |f| = 30\text{-cm at } x = 90 \text{ cm}$$
$$\text{convex lens with } |f| = 10\text{-cm at } x = 110 \text{ cm}$$
(A) Find the location of the final image. (B) Find the size of the final image. (C) Describe the orientation and character of the final image. (D) Compute the magnification of the system.

Volume 2: Waves, Electricity & Magnetism, and Optics

*9. **Fuzzy Panda II**. The absolute value of the focal length of a concave lens is 40 cm. A 12-cm tall fuzzy panda is placed 120 cm before the lens. (A) Draw a ray diagram clearly showing the formation of the image via the three standard rays. Include a legend to indicate how the diagram is scaled. (B) Find the location, orientation, magnification, size, and character of the image.

10. **Monkeyfication I**. A 70-cm tall monkey stands 5 m away from a convex lens with 3-m focal length. (A) Use equations to compute the position of the image. (B) Use equations to compute the height of the image. (C) Compute the magnification. (D) Is the image upright or inverted? (E) Is the image real or virtual?

11. **Monkeyfication II**. Repeat the previous problem if the monkey is now standing one meter from the (same) lens.

12. **Monkeyfication III**. Repeat Monkeyfication I if the lens is concave. (The monkey is 5 m away from the lens.)

*13. **Monkeyfication IV**. Repeat Monkeyfication I using an accurately drawn ray diagram (instead of using the lens equation). Be sure to indicate your scale.

*14. **Monkeyfier**. (A) How far away from a monkey photo should a magnifying glass with 10-cm focal length be placed to produce a real image of the monkey 25 cm from the lens? (B) What will be the magnification?

*15. **Monkey Projector**. A convex lens is to be used to project light onto a screen 40-m away from the lens. Where should a slide 3-cm in width be placed to produce an image on the screen that is 15-m wide?

*16. **Miniature Temple**. A lens system is described below.
   12-cm high object at x = 20 cm
   convex lens with $|f|$ = 30-cm focal length at x = 60 cm
   concave lens with $|f|$ = 40-cm focal length at x = 80 cm
Locate the final image to unlock the miniature temple. (A) Where is the final image located? (B) What is the size and orientation of the final image?

*17. **Red Rubies**. Do you dare to steal the red rubies from the eyes of the pharaoh's statue? Better not – the pyramid would collapse in an instant. Instead, solve this puzzle. An object and a screen are one meter apart. A lens is placed between the object and screen until a focused image appears on the screen that is three times larger than the object. What is the focal length of the lens?

*18. **Ram's Head Amulet**. A 10-cm tall object is 60 cm before a lens with −90 cm focal length. Locate the final image via a ray diagram in order to unlock the ram's head amulet. (A) Draw a large, accurate ray diagram to scale. Draw the three traditional rays to locate the image. (B) Measure the image distance. (C) Measure the image height.

*19. **Purple People-Eater**. The focal length of a convex lens is $|f| = 60$ cm. A 12-cm tall purple people eater is placed 40 cm before the lens. (A) Draw a diagram showing the object, lens, and focus. (B) Draw the traditional three rays on your diagram above to locate the image. Draw and label the image on your diagram above. (C) Use equations to compute the image distance. (D) Use equations to compute the image height. (E) Is the image upright or inverted? (F) Is the image real or virtual? (G) Is the image smaller or larger than the object?

*20. **Monkey Love**. A monkey standing before a lens falls in love with his image. In the diagram below, 1 scaled cm corresponds to 10 cm.

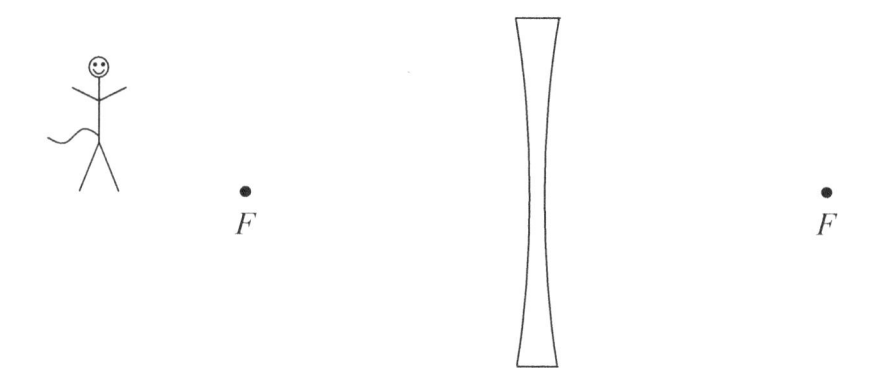

(A) Draw the principal axis. (B) Draw the three traditional rays to locate the image. Draw and label the image on the diagram. (C) What is the object distance? (D) What is the object height? (E) What is the focal length of the lens? (F) What is the image distance? (G) What is the image height? (H) What is the magnification? (I) What is the character of the image? (J) What is the orientation of the image?

*21. **Monkey Image**. As illustrated below, a monkey stands before a lens system.

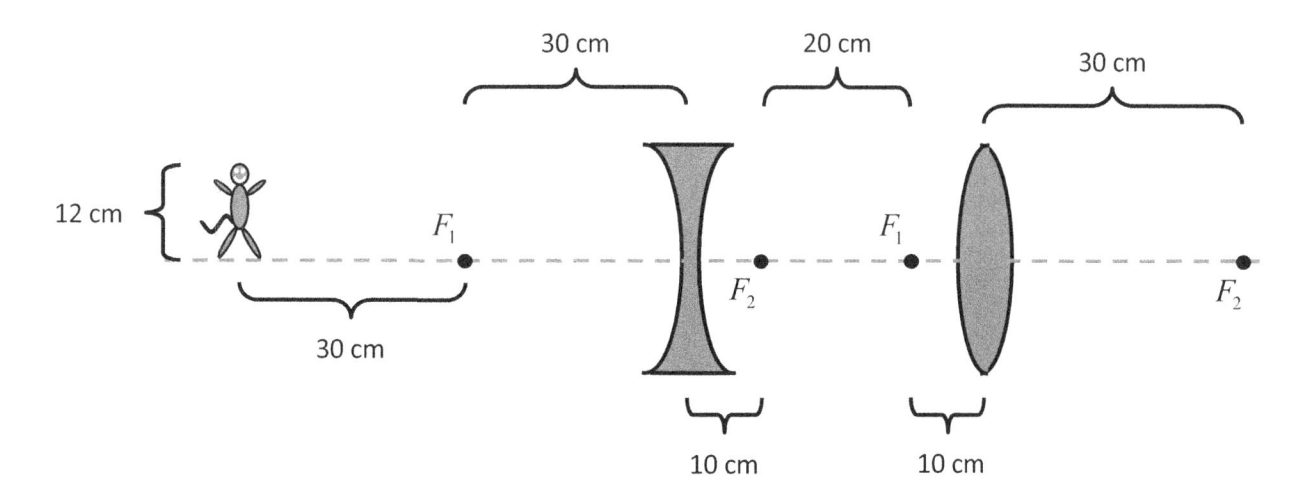

Determine the location, height, magnification, character, and orientation of the final image.

*22. **Smile!** Consider the lens system described below:
>the object is a 20-cm tall smiley face at $x = 200$ cm
>a concave lens with $|f| = 40$ cm at $x = 80$ cm
>a convex lens with $|f| = 100$ cm at $x = 30$ cm

(A) Make a sketch, showing the location of the object and lenses (but don't draw a ray diagram). (B) Find the location, size, magnification, character, and orientation of the final image. When you know where the final image is, draw and label it on your sketch above.

9 Interference and Diffraction

1. **Monkey Interference**. A young monkey performs the double-slit experiment using laser light with 650-nm wavelength. The centers of the slits are separated by 0.05 mm. The screen is 2-m away from the slit. (A) What is the distance from the central bright fringe to the first bright fringe? (B) What is the distance from the central bright fringe to the fifth dark fringe?

2. **Grating**. A monkey shines laser light with 650-nm wavelength upon a diffraction grating. The first-order maximum is located 38º from each side of the central maximum. (A) What is the spacing between two adjacent grating lines? (B) How many lines per centimeter does the grating have?

*3. **Path length**. A monkey places two glass plates such that they are parallel and separated by 0.5 mm. What is the optical path length between the plates when water fills the gap between the plates?

*4. **Vacuum Cell**. A monkey shines laser light with 632.8 nm wavelength toward the two parallel plates shown below. (The angle of the reflected rays is quite exaggerated.) The region between the plates is initially a vacuum cell. Bright reflection is observed by a telescope.

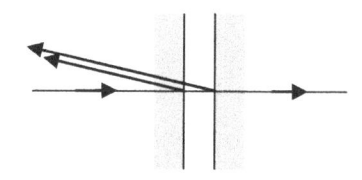

(A) Determine the order number n for the bright fringe that is viewed through the telescope. The two plates are separated by 3 meters. While ammonia is slowly let into the region between the plates (so it is no longer a vacuum), the reflection viewed through the telescope switches from bright to dark to bright 3565 times. That is, 3565 fringes pass through the field of view. (B) What is the order number n after the region between the plates is completely filled with ammonia? [Hint: It will be necessary to use your answer from part (A).] (C) Use your answer to (B) to determine the index of refraction of ammonia.

*5. **Monkey Choice Challenge**. A monkey who spent his life studying optics bets you a bunch of bananas that you can't pass his multiple choice quiz.
A. If the width of a single slit were increased, the separation between fringes would be
 (a) smaller
 (b) unchanged
 (c) greater
B. The location of the first missing order for a multi-slit interference pattern depends on the
 (a) slit width
 (b) slit spacing
 (c) number of slits

C. If Young's double-slit experiment were performed underwater (instead of in air), the separation between fringes would be
- (a) smaller
- (b) unchanged
- (c) greater
- (d) none – there wouldn't be any fringes

D. If the distance between the slit and screen were increased in the double-slit experiment, the separation between fringes would be
- (a) smaller
- (b) unchanged
- (c) greater

E. If blue light were used instead of red light in the double-slit experiment, the separation between fringes would be
- (a) smaller
- (b) unchanged
- (c) greater
- (d) none – there wouldn't be any fringes

F. Diffraction is more pronounced through relatively
- (a) small openings
- (b) large openings
- (c) neither – it's the same for each

G. Waves diffract more when their wavelength is
- (a) short
- (b) long
- (c) neither – both diffract the same

H. Consider plane waves incident upon a barrier with a small opening. After passing through the opening, the waves
- (a) continue as plane waves
- (b) fan out
- (c) converge
- (d) become polarized
- (e) all of these

I. Colors seen when gasoline forms a thin film on water are a demonstration of
- (a) refraction
- (b) reflection
- (c) dispersion
- (d) polarization
- (e) interference

J. Light from a source passes through a lens and forms an image on the other side. Would red light and blue light from the same source produce images at the same place?
- (a) yes
- (b) no – blue light would make an image closer to the lens
- (c) no – red light would make an image closer to the lens

Review 4: Optics

Ape-tics

*1. **Banana Laser**. A laser produces light with 535-nm wavelength in air. The power of the beam is 200 mW and the diameter of the beam is 1.2 mm. (A) What is the frequency of this beam in air? (B) What is the intensity of the beam in air? (C) Suppose the beam enters crown glass. What is the speed of the beam in crown glass? (D) What is the wavelength of the beam in crown glass?

*2. **Banana Laser II**. Consider the diagram below. A red laser beam is incident along path A. When the glass cube is present, light emerges along path B. When the glass cube is removed, light continues along path C.

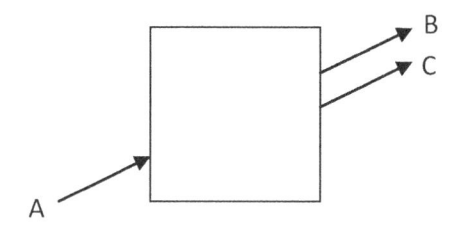

(A) Measure the angle of incidence. (B) Determine the index of refraction of the glass.

3. **Banana Laser III**. Using a diffraction grating with 23,700 lines/inch, the second-order maximum is found at an angle of 72º on each side of the straight-through beam. (A) What is the wavelength of the light? (B) Will the third-order maximum be visible?

4. **Ape-rition I**. A statue of an ape 10-cm tall is placed 60 cm before a convex mirror with 80-cm radius of curvature. (A) Where is the image located? (B) What is the image height? (C) What is the magnification? (D) Is the image real or virtual? (E) Is the image upright or inverted? (F) Is the image enlarged or reduced?

*5. **Ape-rition II**. An ape stands before a curved mirror as illustrated on the following page. The C designates the center of curvature. The diagram is drawn to scale. Measuring devices will be needed to answer these questions. (A) Identify the position of the focus for this mirror on the diagram. (B) Draw the three traditional rays to locate the image on the diagram. (C) Draw and label the image on the diagram. (D) Use your ray diagram to determine the image position. (E) Use your ray diagram to determine the image height. (F) Is the image real or virtual? (G) Is the image upright or inverted? (H) Determine the magnification of the mirror.

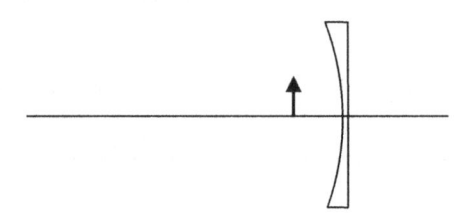

Volume 2: Waves, Electricity & Magnetism, and Optics

*6. **Ape-rition III**. A metal sphere 40-cm in diameter has a reflective surface. An ape's face is 45-cm from the center of the ball. (A) Locate the image of the ape's face. (B) What is the magnification of the ape's face? (C) Is the image of the ape's face real or virtual? (D) Is the image of the ape's face upright or inverted? (E) Is the image of the ape's face enlarged or reduced?

*7. **Ape-rition IV**. An ape is placed before a lens as illustrated below. The F designates one focus. The diagram is drawn to scale. Measuring devices will be needed to answer these questions. (A) Identify the position of the other focus for this lens on the diagram. (B) Draw the three traditional rays to locate the image on the diagram. (C) Draw and label the image on the diagram. (D) Use your ray diagram to measure the image position. (E) Use your ray diagram to measure the image height. Now measure p (and what else?) and use the lens equation to compute q. [Hint: Do your answers to (D) and (F) agree?] (G) Is the image real or virtual? (H) Is the image upright or inverted? (I) Determine the magnification.

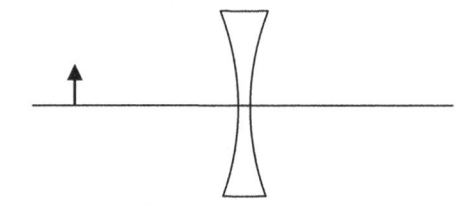

Final Review: Waves, E&M, and Optics

Physics Is History!

*1. **Speed of Sound in Water.** In the 1820's, Jean Daniel Colladon simultaneously triggered his stopwatch and fired a cannon at one end of Lake Geneva. Immediately upon hearing the cannon fire, Jacob Carl Franz Sturm – who stood at the opposite end of the lake (13,487 meters away) – placed a bell underwater and struck it with a hammer. Jean Daniel Colladon, who had a specially prepared ear trumpet dipped in the water at his end of the lake, stopped his stopwatch when it registered the sound of the bell. The stopwatch reading was 0h 0m 48.29s. Use the results of this experiment along with the known speed of sound in air to determine the speed of sound in water.

*2. **Cathode Rays.** In 1897, Nobel laureate Joseph John Thomson (no 'p'!) discovered the electron with an experiment similar to the one illustrated below, in which an electron is heading to the left with an initial speed of 80 cm/s. The electron has a charge of $+1.60 \times 10^{-19}$ C and mass of 9.11×10^{-31} kg. The potential difference between the plates is 1.75×10^{-13} V. Neglect gravity.

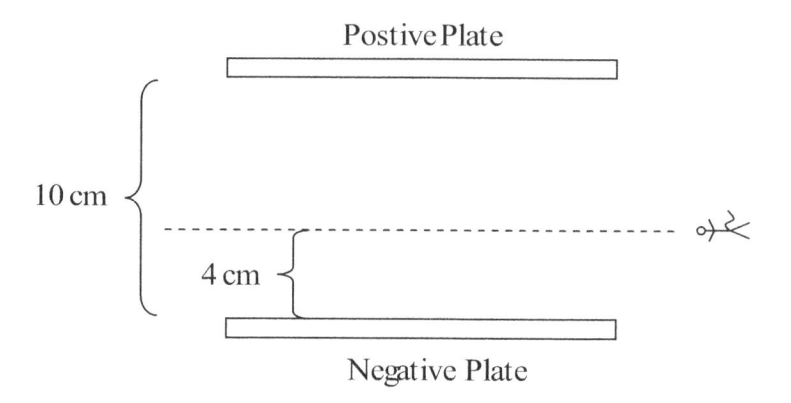

(A) What will be the path of the electron? (That is, which way will it curve, and what kind of curve will it be?) (B) Draw a free-body diagram for the electron. (C) Solve for the electron's acceleration. (D) How far will the electron travel horizontally before striking a plate?

Volume 2: Waves, Electricity & Magnetism, and Optics

*3. **Snell's Law**. Willebrord Snell deduced the law of refraction experimentally. He stated the law inconveniently in terms of the ratio of the cosecants of the angles of incidence and refraction. As illustrated below, a beam of light in zircon is incident upon fluorite. The index of refraction is 2 for zircon and $\sqrt{2}$ for fluorite.

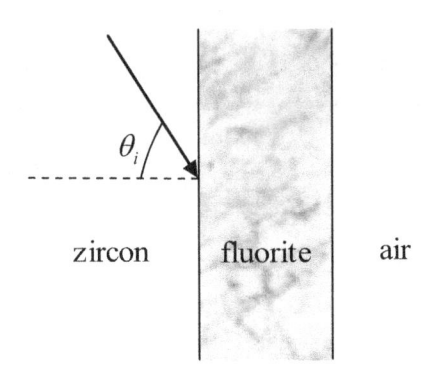

(A) For what range of angles will the beam enter fluorite but not air? (B) Repeat the problem if the order of zircon and fluorite is reversed.

*4. **Solar Death Ray**. In the 3rd century B.C., Archimedes' is fabled to have used a solar death ray – actually, just a large curved mirror – to defend Syracuse against Roman attack. The story is that Archimedes reflected the sun's rays to set fire to the ships when they came within bowshot. (A) Would Archimedes have used a concave or convex mirror? Explain. (B) What is the object in this case? Compared to 10 m, approximately what is the object distance? (C) Draw and label a ray diagram. (Should you draw and label an object or draw incident parallel rays of light? Consider the previous question.) (D) What should be the radius of curvature of a concave spherical mirror in order to harness rays of sunlight to burn an object 10 meters away? (E) If Archimedes instead constructed a convex lens to serve as the solar death ray, what should be the focal length of the convex lens in order to set fire to an object 10 meters away? Draw and label a ray diagram and justify your answer.

*5. **Lens System**. In 1668, Sir Isaac Newton designed and built his own reflecting telescope. A larger version, preserved in the library of the Royal Society, bears the inscription, "Invented by Sir Isaac Newton and made with his own hands, 1671." A lens system is described below.
    12-cm high banana at $x = 20$ cm
    convex lens with $|f| = 30$-cm focal length at $x = 60$ cm
    convex mirror with $|f| = 50$-cm focal length at $x = 80$ cm
Find the location, size, magnification, orientation, and character of the <u>second</u> image.

*6. **Interference**. Thomas Young conducted numerous experiments revealing the wave nature of light. Consider a monochromatic source, such as a laser (which Young did not have, of course), with 500-nm wavelength incident upon one of the following: a single slit, a double slit, a multi-slit, or a diffraction grating. A trace of the interference pattern that appears on the screen is illustrated on the following page.

2.0 cm

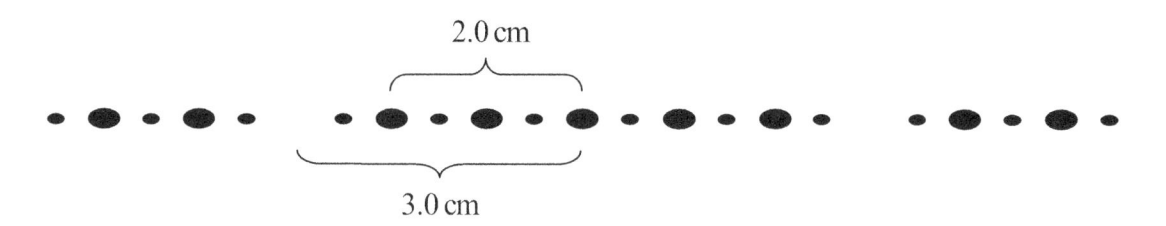

3.0 cm

The experimental setup is illustrated below.

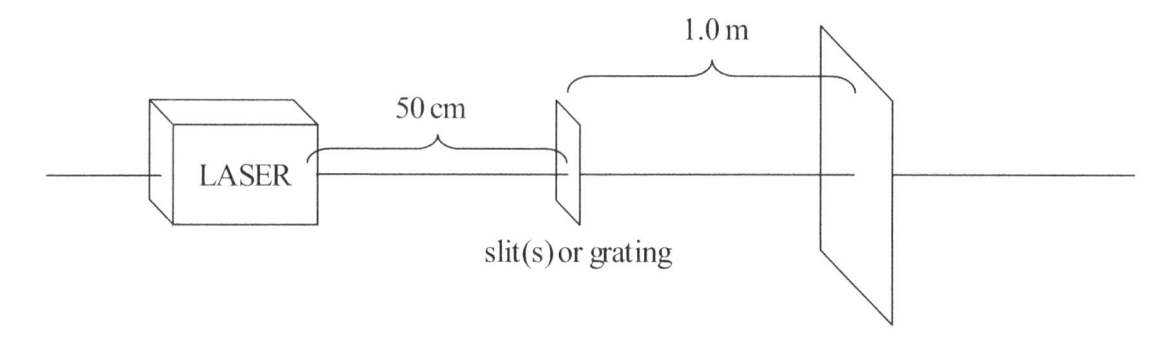

1.0 m

50 cm

LASER

slit(s) or grating

(A) Was the interference pattern produced by a single slit, a double slit, a multi-slit, or a diffraction grating? (B) Determine the relevant slit or grating parameter(s). (If you think it is a single slit find the slit width. If you think it is a double slit, find the slit spacing and the slit width. If you think it is a multi-slit, find the number of slits, the slit spacing, and the slit width. If you think it is a diffraction grating, find the spacing between the lines.)

Volume 2: Waves, Electricity & Magnetism, and Optics

Scholarly Hogwash

*7. **Equivalent Knowledge**. According to the theory of collaborative knowledge:
  - The knowledge (K) of a thinker is a quantity that can be measured in units of whats.
  - The question mark symbol (?) is used to represent units of whats.
  - In a schematic diagram, knowledge is represented as lightning $\left(\rightarrow\!\!\!\!\!\text{⚡}\!-\right)$.
  - The equivalent knowledge of a group of N parallel thinkers is obtained from
  $$\frac{1}{K_p} = \sum_{i=1}^{N} \frac{1}{K_i}$$
  - The equivalent knowledge of a group of N series thinkers is obtained from
  $$K_s = \sum_{i=1}^{N} K_i$$

A group of thinkers is collaborating on a project as illustrated in the schematic diagram below.

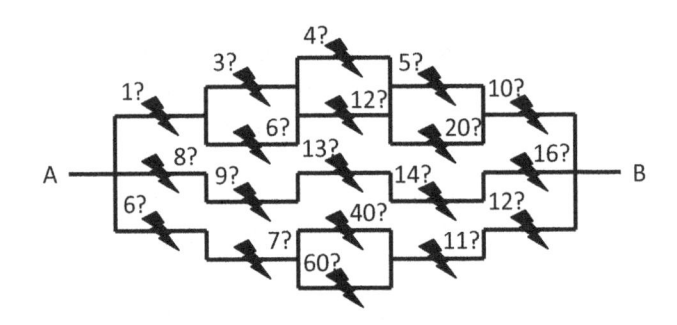

(A) Do collaborative thinkers behave more like capacitors or resistors? Explain. (B) Find the equivalent knowledge of the collaboration between points A and B.

*8. **Earned Salary**. According to the law of earned salaries:
  - A worker's salary is represented by the symbol M, which is measured in dollars ($).
  - Knowledge is power and time is money. Since power equals work over time, it follows that
  $$W = KM$$
  - A company is defined as a device that provides work to a group of workers.
  - In a schematic diagram, a company is represented as a sun:
  - Salary is the same for each member of a group thinking in series.
  - Work is the same for each member of a group thinking in parallel.
  - Knowledge is computed according to the theory of collaborative knowledge.

A group of thinkers is collaborating on a project as illustrated in the schematic diagram that follows.

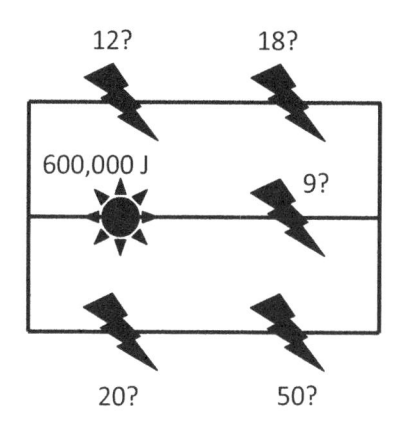

(A) How are ?'s related to $'s? (B) Determine the salary for each worker in the collaboration.

*9. **Animal Magnetism**. According to the law of animal magnetism:
- A moving emotional particles creates an animal magnetic field $(\vec{A})$.
- Animal magnetic fields are measured in units of Ardor $(\forall)$.
- An emotional particle traveling in a uniform animal magnetic field $\vec{A}$ experiences a force according to the equation $F = p\upsilon A \sin\theta$, where θ is the angle between the velocity and the animal magnetic field.
- Emotional particles carry passion, as described by Cool Hombrè's law (see p. 42).
- The same right-hand rules of physics apply to the direction of animal magnetic fields.

(A) An abhoron with a passion of −20 m♥ and a mass of 4 g travels in a circle with a 6-m diameter with constant speed in a uniform animal magnetic field of $0.5\,\forall$. What is its speed?

*10. **Brain Waves**. According to the theory of brain waves:
- The process of thinking results in the transmission of brain waves.
- Brain waves feature a characteristic amplitude, frequency, and wavelength.
- The frequency and wavelength of brain waves are related in the same way as they are for all other waves.
- All brain waves travel the same speed, 50 m/s.

Proponents of the brain wave theory purport that people are much smarter when they think in the shower provided that they produce brain waves at one of the resonance frequencies.

Consider a man singing in an outdoor shower that is 2-m tall, 1-m wide, and 1-m deep. The shower has four walls, but no roof. Consult the diagram on the following page. (A) Draw the fundamental and first overtone separately for standing waves produced along the height and width. (B) Determine the resonance frequencies for the fundamental and first overtone separately for standing brain waves produced along the height and width of the shower.

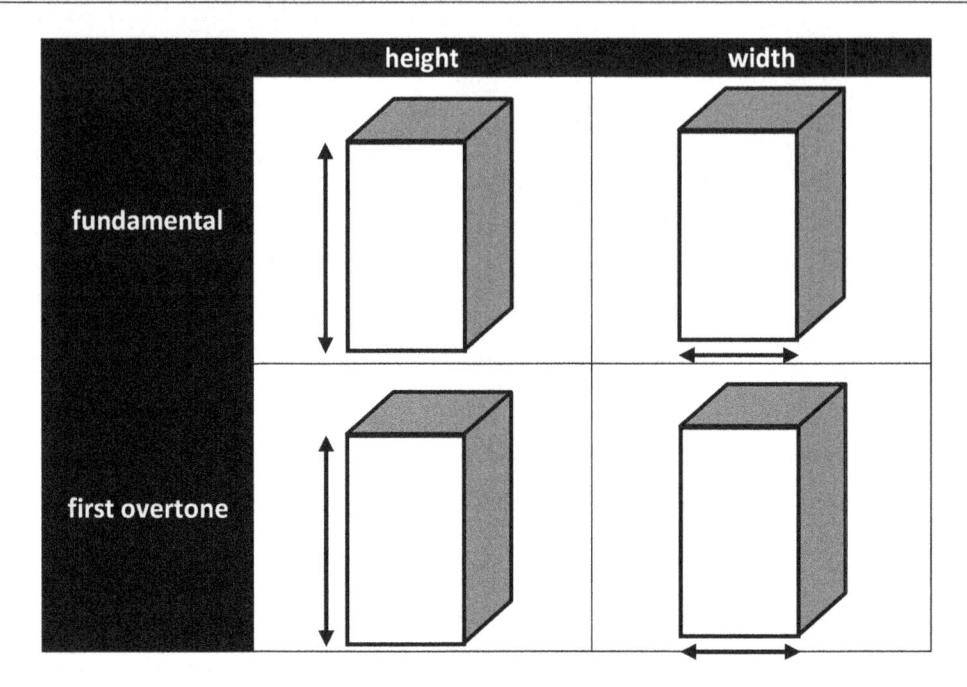

Volume 2: Waves, Electricity & Magnetism, and Optics

	height	width
fundamental		
first overtone		

*11. **Rainbow Formation**. According to the theory of magnetic dispersion:
- During precipitation, numerous raindrops develop a net electric charge by gaining or losing one or more electrons.
- These accelerating charged raindrops emit photons.
- The frequencies of the emitted photons are proportional to the speeds of the raindrops.
- The raindrops at higher altitudes have lower speeds and therefore lower frequencies and higher wavelengths. Thus, red is seen at the top of the rainbow.

The raindrops at lower altitudes have higher speeds and therefore higher frequencies and lower wavelengths. Thus, violet is seen at the bottom of the rainbow. (A) Can the theory of magnetic dispersion account for the formation of the secondary rainbow? Explain. (B) Draw the path that a ray of light follows in a raindrop in the formation of the primary rainbow according to the accepted theory of physics (i.e. not magnetic dispersion). (C) Describe how your diagram explains which color appears at the top of the primary rainbow. (D) Draw and label a diagram that shows where an observer should stand relative to the sun and rain in order to observe the primary rainbow. (E) Draw the path that a ray of light follows in a raindrop in the formation of the secondary rainbow according to the accepted theory of physics. (F) Describe how your diagram explains which color appears at the top of the secondary rainbow. Also, explain why the secondary appears weaker than the primary.

*12. **Refraction**. According to the theory of polarized spins:
- The atoms of a glass are polarized such that their spins tend to be perpendicular to the surface of the glass.
- When a ray of light in air enters the glass, the atomic spins influence the photons such that the beam of light refracts through the glass.

(A) Can the theory of polarized spins account for the refraction of light passing from glass to air? Explain. (B) As illustrated on the following page, a ray of light in air is incident upon a solid glass cube at an angle of 45º with respect to the normal. The glass cube has $\sqrt{2}$-cm sides and an index of refraction of $\sqrt{2}$. According to the accepted theory of physics (i.e. not polarized spins), when the glass cube is lifted out of the path of the laser beam, how far does the beam shift?

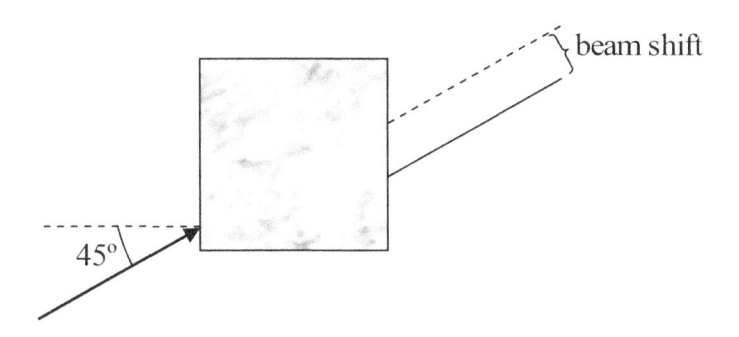

*13. **Mood Meters**. According to the theory of mood meter design:
- People radiate tiny particles called gleeons (γ).
- People radiate more gleeons when they are happy and fewer when they are sad.
- Gleeflow (Γ) is the rate of flow of gleeons, measured in units of flowers (φ).
- Glee potential difference ($\Delta\Psi$) is a measure of how much work is required to move a gleeon between two points, measured in units of jolts (j).
- Gleeresistance (ρ) is a measure of the opposition to gleeflow, with units of arms (α).
- Gleeons obey the equation $\Delta\Psi = \Gamma\rho$.
- A gleemeter measures gleeflow, but has a significant internal gleeresistance.
- A flometer measures gleeflow, which has a relatively small internal gleeresistance.
- A joltmeter measures glee potential difference between two points.

Consider a gleemeter with a 20-α internal gleeresistance that deflects full-scale for a 50-mφ gleecurrent. (A) How large a gleeresistance must be connected in parallel with such a gleemeter in order to construct a 5-φ flometer from it? (B) How large a gleeresistance must be connected in series with such a gleemeter in order to construct a 4-j joltmeter from it?

Answers to Selected Problems

Chapter 1: Simple Harmonic Motion

1. **Burning Coals**.
(A) 15 cm
(B) 3.0 sec
(C) 0.33 Hz
(D) 2.1 rad/s

3. **Physics Warms the Heart**.
(A) 310 K
(B) 15 K
(C) 1.8 sec
(D) 0.56 Hz
(E) 3.5 rad/s
(F) -140º
(G)
$$T(t) = 15[\text{K}]\sin\left(3.5[\text{rad/s}]\,t - \frac{7\pi}{9}[\text{rad}]\right) + 310[\text{K}]$$

5. **Simple Harmonkey Motion**.
(A) ±3.0 m
(B) 16 sec
(C) -1.7 m
(D) 0.99 m/s
(E) 0.27 m/s^2

6. **Shish Kabob Seconds**.
(A) -0.96 m, -1.00 m, 0.0045 N/m, 3.07 × 10^{-9} J (potential for large round-off error), -0.0018 m/s^2.
(C) $9.88 + 20.9n$ or $16.9 + 20.9n$ sec, where n is an integer.

7. **Springtime**.
(A) 4.11 Hz
(B) π/2 rad
(C) 1.46 m/s
(D) 37.0 m/s^2

13. **Cuckoo! Cuckoo!**
(A) 10.7 N/cm
(B) 26.7 cm/s, 88.8 cm/s^2

14. **Good Morning, Physics!**
(A) 4.87 m/s
(B) 0.234 sec
(C) $a(t) = -16\pi \cos(4\sqrt{5}\,[\text{rad/s}]t)$ [m/s^2]
(D) -1.99 m/s

16. **Penguilum**. 11

17. **Planet Ribbit**. 0.19 N

20. **Snow SHO**.
(A) -9.42º
(B) 1.22 rad/s
(C) 4.28 rad/s^2
(D) 0.863 m/s
(E) 299 ms

23. **Monkeystick**. 1.04 s

Chapter 2: Waves

1. **Surfer Physics**.
(A) 12 m
(B) 50 m
(C) 0.133 Hz
(D) 7.5 sec
(E) 6.67 m/s

5. **Dolphin Symphony**.
(A) 30 cm
(B) 7.0 cm

6. **Meet Newton**. 319 m/s

7. **Dogarithm**. (A) 6.31×10^{-5} W/m^2. (B) 81.0 dB.

8. **Physics Causes Nightmares!** 190 m/s

9. **Doggler Effect**. $0.74 \times$

12. **Shark Wave**. 4.61 sec

13. **Physics Causes Migraines!** 725 m/s

14. **Sonic Doom**. "Mach 1.34"

15. **Frankenstein's Lab**. 33.7 Hz, 67.4 Hz, 101 Hz

18. **You've Flipped Your Lid!** Neon

Review 1: Oscillations and Waves

7. **Cow**.
(A) 61 m/s
(B) 3.55 km

8. **Monkey**.
(A) 228 osc./min.
(B) ±10.5 cm
(C) $-68.6 \sin\left(23.9t + \dfrac{\pi}{2}\right)$

9. **Owl**.
(A) 19.4 m/s
(C) 32.3 Hz, 64.5 Hz, 96.8 Hz

10. **Sheep**.
(A) 3.22 m
(B) -222º
(C) 1.18 m/s

11. **Parrot**.
(A) 0.045 N/m
(B) 3.28 s

Chapter 3: Electric Field

5. **Prize-Winning Apples**. 2.32×10^4 kg

6. **Apples and Oranges**. 6.29 N, 0.506 N

10. **Triapple**. 82.5 N, 167º.

11. **Apples and Bananas**.
(A) $(4.29 \times 10^6 \text{ N/C}, 0°)$
(B) $(6.80 \times 10^4 \text{ N/C}, 47.0°)$
(C) -1.04 m

14. **A Rose that Froze**. 22.2º, 0.343 N

15. **Mosquito Physics**. 6.50×10^7 N/C, 175º; 1.95 kN

17. **Gauss's Claw I**. $\vec{E}(r_c) = \dfrac{2k\lambda}{r_c}\hat{r}_c$

18. **Gauss's Claw II**.
$$\vec{E}\left(x \leq -\dfrac{d}{2}\right) = -\dfrac{\rho d}{2\varepsilon_0}\hat{i}$$
$$\vec{E}\left(-\dfrac{d}{2} \leq x \leq \dfrac{d}{2}\right) = \dfrac{\rho x}{\varepsilon_0}\hat{i}$$
$$\vec{E}\left(\dfrac{d}{2} \leq x\right) = \dfrac{\rho d}{2\varepsilon_0}\hat{i}$$

19. **Gauss's Claw III**.
(A) $-2Q$
(B) $5Q$
(C) $Q/4$

Volume 2: Waves, Electricity & Magnetism, and Optics

(D)
$$\vec{E}(r \le R_1) = \frac{2kQr}{R_1^3}\hat{r}$$
$$\vec{E}(R_1 \le r \le R_2) = \frac{2kQ}{r^2}\hat{r}$$
$$\vec{E}(R_2 \le r \le R_3) = 0$$
$$\vec{E}(R_3 \le r) = \frac{5kQ}{r^2}\hat{r}$$

20. **Square.** 1.59×10^7 V

21. **Ca-pie-citor I.** $0.00131c$.

22. **Ca-pie-citor II.** 7.30 cm.

Chapter 4: DC Circuits

1. **Banana Capacitors.** 20 µF, 30 µF

3. **Smiley Face.**
(A) 24.0 nF
(B) 43.2 V
(C) 1.73 µC
(D) 8.96 µJ
(E) 0.691 µC

4. **Home Sweet Home.**
(A) 25.0 µF
(B) 0.750 V
(C) 7.50 µC
(D) 0.900 µJ
(E) 4.50 µC

8. **Permit This.**
Bottom Right: 60.2 pF (10.0 pF)
Botttom Left: 5.41 pF (7.43 pF, 19.8 pF)
Top: 5.38 pF (3.72 pF, 6.55 pF, 3.01 pF)
(A) 10.3 pF
(B) 265 pJ
(C) 482 nC

9. **Resist This.** 3.45 Ω

10. **Resistor Journey.**
(A) 44.7 Ω
(B) 35.3 Ω

11. **The force that's not a force, of course.**
13.2 V, 10.0 Ω

12. **I Scream for Physics!**
(A) 120 Ω
(B) 800 mA
(C) 43.2 W
(D) 24.0 V
(E) 600 mA
(F) 108 V

13. **Physics in a Bottle.**
(A) 36.0 Ω
(B) 2.00 A
(C) 853 mW
(D) 667 mA
(E) 16.0 V

16. **Sugar Cube.** $\frac{7}{12}\Omega$

17. **Resistor Cube.** $\frac{3R}{4}$

18. **Frosty the Circuit.**
(A) 150 mA
(B) 1.5 V
(C) 180 mW
(D) top
(E) c, b, a, d

19. **Monkey Physics.**
(A) 61/177 A
(B) -922/177 V
(C) 35287/31329 W
(D) bottom
(E) b

84

Review 2: Electricity

5. **Reseastor Bunny.** $24\,\Omega, 1.6\,\text{A}, 80.6\,\text{W}$

6. **Cabasketor.** $16\,\mu\text{F}, 960\,\mu\text{C}, 20\,\text{mJ}$

7. **Reseastor Egg.** $\dfrac{5}{9}\,\text{A}, \dfrac{62}{27}\,\text{A}, \dfrac{122}{27}\,\text{V}$

Chapter 5: Magnetic Field

1. **Which Way's Right?**

⊙ ↙ ↙ ⊙ 0 ↖

⊗ ⊗ ⊙ ↓ ← →

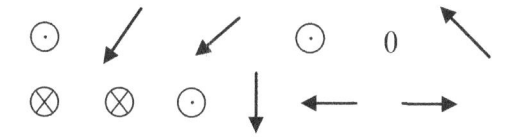

3. **Monknetic Force I.**
(A) 0.408 N, south
(B) 0.394 N

5. **Monkey Tails I.**
2.88×10^{-8} N, right

6. **Monkey Tails II.** 0.0012 Nm
Assuming no pivots, it will rotate about an axis passing through its center of mass.

7. **Monknetic Force II.**
(A) 1.20×10^{-16} N, north. (B) 1.75 μm.

9. **Ampère's Claw.**

(A) $\vec{B}(r \le R_0) = \dfrac{\mu_0 I_0 r}{2\pi R_0^2}\hat{\boldsymbol{\theta}}$

$\vec{B}(R_0 \le r \le 2R_0) = \dfrac{\mu_0 I_0}{2\pi r}\hat{\boldsymbol{\theta}}$

$\vec{B}(2R_0 \le r \le 3R_0) = \dfrac{\mu_0 I_0 (9R_0^2 - r^2)}{10\pi r R_0^2}\hat{\boldsymbol{\theta}}$

$\vec{B}(3R_0 \le r) = 0$

(B) $\dfrac{I_0}{4}$

10. **Love Meter.**
(A) 50.0 mΩ
(B) 950 Ω

Chapter 6: Faraday's Law

1. **Which Way's Right?** cw, ccw, cw, ccw, cw, ccw, ccw, cw, ccw, 0, ccw, cw, ccw, ccw, cw

5. **Monkeying Around.** 3.01 A, cw

6. **Slide a Rod.** 5.38 m/s

7. **Shake a Snake.**
(A) 0.264 Ω
(B) 0.0528 V
(C) 19.0 ms

9. **Electrocute an Eel.**
(A) 5
(B) 120 mA is the naïve answer. The correct method is more involved, accounting for the effects of mutual inductance and ensuring conservation of energy.
(C) 1.20 kV

Review 3: Magnetism

5. **Gorilla.** 1.75×10^{-8} N

7. **Chimp.** 6.0 T

8. **Gibbon.** 0.36 Nm

9. **Tarsier.** 1.06×10^{-4} Tm^2
Chimp. 177 mA
Gibbon. 225

Chapter 7: Snell's Law

2. **Laser Tag I.** 20.3º

3. **Laser Tag II.** (A) 44.5º (B) 59.1º

Volume 2: Waves, Electricity & Magnetism, and Optics

6. **Beam Shift**. 1.55 cm (If you are a little off, you probably found the wrong side, not realizing that the actual side you need is a little different.)

Chapter 8: Mirrors and Lenses

3. **Scarab Necklace**. 160º

4. **Bunch of Bananas**.
(A) -1.88 m , 26.3 cm , 0.375×
(B) -1.50 m , 105 cm , 1.50×

6. **Miniature Monkey**. 80 cm

8. **Lens System**. 123 cm, -0.783 cm, inverted, real, -0.130×

10. **Monkeyfication I**. 7.50 cm, -1.05 cm

11. **Monkeyfication II**. -1.50 cm, 1.05 cm

12. **Monkeyfication III**. -1.88 cm, 0.263 cm

Chapter 9: Interference and Diffraction

1. **Monkey Interference**. (A) 2.6 cm (B) 11.5 cm

2. **Grating**. (A) 1.06 µm (B) 9,470 lines/cm

Review 4: Optics

3. **Banana Laser III**.
(A) 510 nm

4. **Ape-rition I**.
(A) -24 cm

Made in United States
Orlando, FL
19 November 2022

24755069R00050